Praise for

Essential Rainwater Harvesting

This book is essential for anyone intending to live a healthy sustainable lifestyle. Rob and Michelle Avis are brilliant permaculture designers of the highest caliber worldwide. Their work is always accurate, precise, and detailed, and this book is exactly that. It will probably be in print for the next century. Buy one for yourself and it will be useful for all future generations.

— Geoff Lawton, Permaculture consultant, designer and teacher, www.geofflawtononline.com

Essential Rainwater Harvesting is chock-full of detailed advice for how to design a personalized rainwater harvesting system for a wide variety of purposes. The authors have many years of professional experience to draw on, giving you all you need to know to accomplish a successful plan. I wish I had this book when I built my first harvesting system.

— Kelly Hart, author, *Essential Earthbag Construction*
www.naturalbuildingblog.com

Rapid and massive climate change in the decades ahead will strain water resources in unprecedented ways throughout the world. You owe it to your family to follow the prescriptions in this book and be prepared for droughts and shortages through rainwater harvesting. You'll learn everything you need to know to make it happen by following "the Avis method."

— Jerry Yudelson, P.E., author of *Dry Run: Preventing the Next Urban Water Crisis*

This is going on my shelf as a crucial reference for design work. Accessible and quick to reference, yet as thorough as one could ever need on the subject. A fantastic design tool.

— Ben Falk, author, *The Resilient Farm and Homestead*

With growing populations, fast-depleting aquifers, and climatic instability, the information in *Essential Rainwater Harvesting* will only get more valuable as we move into the future. The science-based approach that Rob and Michelle Avis take should help not only to get your system set up, but also educate policymakers with the most recent research.

— Dave Boehnlein, author, *Practical Permaculture,* and Principal, Terra Phoenix Design

As the world warms, and water supply becomes more and more insecure in many parts of the world, knowing how to safely capture and store water is fast becoming a key skill for anyone wanting to make their lives more resilient. This fine tome tells you everything you need to know in an accessible and inspiring way. Go to it! Become a rainwater harvester! Embrace your downpipes and filtration systems! And all the while with this brilliant book stuffed into your back pocket.

— Rob Hopkins, founder, the Transition movement, transitionnetwork.org

Inspired to save your own rainwater for garden and home? *Essential Rainwater Harvesting* is bursting with personal, practical, and precise information to get gardeners saving water right now. In this comprehensive look at a complex subject, I discovered that everyone benefits from a little extra water sense, so dive in now and start saving water today.

— Donna Balzer, horticulturalist and author, *No Guff Vegetable Gardening*

This book is an essential addition to every homeowner's library. The challenge of designing a safe and reliable rainwater harvesting system for your home is presented in a logical and easy-to-follow way, and makes it easier than ever to take advantage of the benefits of rainwater harvesting for urban and rural sustainability.

— Rene Michalak, Governor, Canadian Association for Rainwater Management, Prairie Chapter

I wish this book and the others in the *Sustainable Building Essentials* series had been available when we built our home years ago. Michelle and Rob do a great job presenting not only rainwater harvesting, but how it fits into a whole-systems approach for increased resiliency. They back up their recommendations with the latest research on safe rainwater utilization.

— Laurie Neverman, Common Sense Home, commonsensehome.com

A great book that appeals to a great variety of audience from the inexperienced to the professional, and covering all aspects of rainwater harvesting from planning to commissioning. It is well-structured, allowing the reader to pick up and implement the basics, or delve into the details with complex analysis on feasibility and optimization. This book is unique with all the tools being provided for the calculations and explained clearly, as well as fantastic advice on best location, materials selection, and plumbing connections. Well done Rob and Michelle; this book certainly compliments the *Sustainable Building Essentials* series.

— Rod Wiese, Managing Director & Principal Engineer, Storm Consulting, stormconsulting.com.au

Essential Rainwater Harvesting is the most holistic and actionable resource on rainwater harvesting that I have encountered. The book covers all the elements of a well-designed rainwater harvesting system in a succinct and easy-to-read format. Using their engineering and permaculture expertise, Rob and Michelle have created an easy-to-follow guide walking the reader through the design process for their rainwater harvesting system. The powerful spreadsheet tool allows users to confidently build a custom system for their specific needs. This book is an excellent resource for professionals and novices alike.

— Andrea Rocchio, Water Harvesting Practitioner

Rob and Michelle have taken a lot of complex information and created a really down-to-earth step-by-step guide. This user-friendly handbook helps you understand and define your own water harvesting goals before digging into the specifics of rainwater harvesting systems. I love the examples, explanations, and personal insights that make data collection and use so approachable!

— Steven Biggs , Radio Host: Farm. Food. Garden

essential
RAINWATER HARVESTING

sustainable
building
essentials

essential
RAINWATER
HARVESTING
a guide to home-scale system design

Rob Avis and Michelle Avis

VERGE
PERMACULTURE

ADAPTIVE HABITAT

new society
PUBLISHERS

New Society
Sustainable Building Essentials Series

Series editors

Chris Magwood and Jen Feigin

Title list

Essential Hempcrete Construction, Chris Magwood

Essential Prefab Straw Bale Construction, Chris Magwood

Essential Building Science, Jacob Deva Racusin

Essential Light Straw Clay Construction, Lydia Doleman

Essential Sustainable Home Design, Chris Magwood

Essential Cordwood Building, Rob Roy

Essential Earthbag Construction, Kelly Hart

Essential Natural Plasters, Michael Henry & Tina Therrien

Essential Composting Toilets, Gord Baird & Ann Baird

Essential Rainwater Harvesting, Rob Avis & Michelle Avis

See www.newsociety.com/SBES for a complete list of new and forthcoming series titles.

THE SUSTAINABLE BUILDING ESSENTIALS SERIES covers the full range of natural and green building techniques with a focus on sustainable materials and methods and code compliance. Firmly rooted in sound building science and drawing on decades of experience, these large-format, highly illustrated manuals deliver comprehensive, practical guidance from leading experts using a well-organized step-by-step approach. Whether your interest is foundations, walls, insulation, mechanical systems, or final finishes, these unique books present the essential information on each topic including:

- Material specifications, testing, and building code references
- Plan drawings for all common applications
- Tool lists and complete installation instructions
- Finishing, maintenance, and renovation techniques
- Budgeting and labor estimates
- Additional resources

Written by the world's leading sustainable builders, designers, and engineers, these succinct, user-friendly handbooks are indispensable tools for any project where accurate and reliable information is key to success. GET THE ESSENTIALS!

Cover design by Diane McIntosh.
Illustrations by Dale Brownson. Images © Verge Permaculture
Interior photos © Verge Permaculture unless otherwise noted.
Sidebar water background: AdobeStock_67829647.

Printed in Canada. First printing November 2018.

This book is intended to be educational and informative. It is not intended to serve as a guide. The author and publisher disclaim all responsibility for any liability, loss, or risk that may be associated with the application of any of the contents of this book.

Inquiries regarding requests to reprint all or part of *Essential Rainwater Harvesting* should be addressed to New Society Publishers at the address below. To order directly from the publishers, please call toll-free (North America) 1-800-567-6772, or order online at www.newsociety.com

Any other inquiries can be directed by mail to:
New Society Publishers
P.O. Box 189, Gabriola Island, BC V0R 1X0, Canada
(250) 247-9737

LIBRARY AND ARCHIVES CANADA CATALOGUING IN PUBLICATION

Avis, Rob, 1979-, author
     Essential rainwater harvesting : a guide to home-scale system design
/ Rob Avis and Michelle Avis.

(Sustainable building essentials)
Includes bibliographical references and index.
Issued in print and electronic formats.
ISBN 978-0-86571-874-6 (softcover).--ISBN 978-1-55092-667-5 (PDF).--
ISBN 978-1-77142-262-8 (EPUB)

     1. Water harvesting. I. Avis, Michelle, 1980-, author II. Title.
III. Title: Rainwater harvesting. IV. Series: Sustainable building essentials

TD418.A95 2018   628.1'42   C2018-905361-5

                                          C2018-905362-3

Funded by the Government of Canada | Financé par le gouvernement du Canada | Canada

New Society Publishers' mission is to publish books that contribute in fundamental ways to building an ecologically sustainable and just society, and to do so with the least possible impact on the environment, in a manner that models this vision.

Contents

To Bill Mollison, Geoff Lawton, and David Holmgren.
Nearly a decade ago, your ideas ignited a spark and initiated the path we are now on.

To our mentors, technical advisors, advocates, and other authors
on rainwater harvesting, particularly Brad Lancaster and Art Ludwig.
An enormous thank you to Peter Coombes of Urban Water Cycle Solutions,
whose contributions, technical review, and perspective was instrumental to this book.

To our incredibly supportive family and friends.
We count ourselves very lucky to be surrounded by such a great community.
A particularly big thank you to Annette (Michelle's mother)
for her never-ending support and willingness to help.

To our children, Rowan and Naomi, who motivate us more than anything else
to continue working hard to make the world a better place.

Foreword

By Peter J. Coombes

WE LIVE IN A SYSTEM where all actions and things are connected. This complex system called Earth includes a biosphere, an atmosphere, oceans, waterways, soil profiles, flora, and fauna that are linked in a feedback, or cybernetic, system. Human systems operate within these natural systems that are altered by our dependence on ecosystem services and natural resources.

Ecological systems contribute to human welfare, both directly and indirectly to represent substantial economic value. Earth systems and human welfare are also impacted by climate processes. We are also challenged by population growth, increasing urbanization, a changing climate, and environmental degradation. Jay Forrester and Donella Meadows have discussed how our Earth and the biosphere that sustain all living things is finite. They asked what comes after growth and highlighted that we have limited time to respond to escalating challenges. Human creativity has improved our prospects.

The 2015 World Economic Forum concluded that water scarcity — a lack of freshwater resources to meet demand — will be the largest global problem over the next decade, affecting over 800 million people who have no access to safe water. This emerging global water crisis involves substantial water shortages as well as the risk of waterborne diseases. Amidst growing social inequality and escalating environmental degradation, it is our responsibility to share scarce resources across multiple generations and species. Market economies alone cannot deliver public goods, equity, and sustainability.

Bill Mollison and David Holmgren's Permaculture concept is based on creating local solutions that influence the wider environment, organic agriculture, appropriate technology, community, and other efforts aimed at creating a sustainable world. A central tenant of the Permaculture movement is to "think globally and act locally," which refers to worldwide environmental improvements being achieved through positive, affordable action taken by local communities. This includes mimicking natural regimes to protect ecosystems. An excellent example of this is harvesting rainfall as close to where it falls for water supply.

Rainwater harvesting is an ancient practice that evolved over 1000s of years in most cultures. Early Egyptian, Indian, Korean, Chinese, and Roman settlements collected and stored rainwater to sustain human settlements, animals, and agriculture. These civilizations — and many others — learned to harvest and store intermittent rainfall from different surfaces, and to manage catchment systems to protect human health. Rainwater harvesting succeeded in drought-prone regions for millennia, and time-tested practices were developed. However, the utility, efficiency, and sustainability of these practices has been mostly forgotten in the recent history of centralized water supplies.

The choice of water supplies to urban areas changed to single-purpose solutions that were incrementally developed at a centralized scale to meet increasing local demands. This new water-supply paradigm improved water security, underpinned economic growth, and generated revenue for governments. The associated improved sanitation provided substantial human health benefits in cities. Rainwater harvesting from roof catchments evolved to become a source of water in rural areas.

Rainwater harvesting is an important source of local water supply in the context of global water and environmental challenge. Nevertheless, the viability and safety of rainwater harvesting is often contested by water and health authorities. Similar to all water supplies, rainwater harvesting systems should be appropriately designed and managed for their intended end uses.

Australia is a substantial case study with over 2.3 million households relying on rainwater for drinking, and more than 6.3 million people use rainwater for some household use. In spite of some claims of widespread health concerns, there have been no health epidemics or widespread illness. David Cunliffe explains in Australia's national ENHealth Guidelines that results of water-quality testing were inconsistent with perceived public health impacts — illness resulting from rainwater supplies is rare. Epidemiological health surveys by Jane Heyworth and Shelly Rodrigo confirmed that rainwater harvesting systems provide similar public health outcomes to mains water supplies. However, there were a small number of situations where poor design, inadequate maintenance, and a lack of understanding about rainwater harvesting led to human illness.

Historian Pat Troy provides useful insight into the dichotomy of opinions about rainwater harvesting. In the 1800s, rainwater collected from roofs and stored in tanks was a primary water source. It was often suggested that coal dust and disease entered tanks via roofs. However, rainwater harvested from roofs and stored above ground caused very few health problems. Rainfall harvesting from trafficable surfaces and stored underground was associated with greater health risks. Legislation was passed in the 1890s to ensure the economic viability of government water utilities by effectively banning use of rainwater, which compelled citizens to pay for mains water. Prior to the 1990s, rainwater harvesting was actively discouraged in urban areas by government water utilities and agencies who claimed rainwater harvesting was illegal and dangerous. The paradigm of centralized water treatment and distribution networks to provide permanent water supplies and protect the health of large populations was inconsistent with an understanding of rainwater supply to a single household. However, rainwater harvesting has again become important in providing household water supply.

A background of living in households dependent on rainwater (similar to most rural Australians) and curiosity about the lack of human illness from Australia's many "untreated" rainwater harvesting systems led me to conduct a national survey of rainwater harvesting. We examined water quality using all available molecular (for example DNA), microbial, and chemical methods at a large number of households over a decade. The quantity of rainwater harvested was also observed using smart meters, social surveys, and local weather stations.

We discovered that water treatment processes of flocculation, settlement, sorption, and bio-reaction naturally operate in rainwater storages to improve the quality of stored rainwater. Rainwater storages are bioreactors with biofilms at the water surface micro layer, on internal walls, and at the bottom, as sludge. (Rainwater should not be drawn from the bottom 100 mm [4 inches] of a rainwater tank to avoid disturbance of the sludge.) Rainwater storages act as balanced ecosystems similar to environmental systems that improve water quality. Pathogens were rarely found in rainwater collected from roofs in above-ground storages. Most rainwater supplies was compliant with the chemical and metal values in the Australian Drinking Water Guidelines. The quality of rainwater varied in response to rainfall inputs, and improved after rainfall.

Most households used a "natural" multiple barrier treatment train where roof, gutters, downpipes, storages, and supply arrangements improved rainwater quality and excluded potential sources of contamination throughout the system. These households were not just relying on traditional end-of-line treatment. Rainwater harvesting systems were providing substantial rainwater yields, even during drought, a finding that was consistent with Brad Lancaster's observations from Arizona in the US. Importantly, these rainwater yields and the correct design solutions were only revealed by detailed and well-informed analysis and modeling.

Systems analysis also revealed that "home-scale" rainwater harvesting provided regional-scale benefits for stormwater management, waterways health, urban amenity, and water security. However, the designer of small-scale rainwater harvesting systems encounters a myriad of mixed messages from authorities, mains water industry, health officials, and product suppliers. In many respects, we are remote from the wisdom of people who successfully operated rainwater harvesting systems for decades and millennia.

It is timely that Michelle and Rob Avis have written an evidence-based practical book, *Essential Rainwater Harvesting*, to assist in the design and management of home-scale rainwater harvesting systems. Michelle and Rob's strong background in Permaculture, their practical experience, and their inquiring minds have ensured that this book is based on the required holistic considerations that are essential for design of a successful rainwater harvesting system. Importantly, they recognize that rainwater harvesting systems are part of a whole-of-society strategy for security, resilience, and sustainability that considers the Earth system that sustains us. It is also recognized that our homes and behaviors are also complex systems that impact on performance of rainwater harvesting systems.

This book cuts through the conflicting messages by providing important discussion of the fundamentals and key components or "anatomy" of rainwater harvesting systems that include collection and pre-filtration, storage, pumps, and ensuring adequate water quality for desired end uses. The reader is also offered support through a website and a tool kit that helps evaluate designs for rainwater yield. This combination of independent guidance fills a void in a confounded marketplace and regulatory environment. Most regulation is focused on mains water industry end-of-line treatment — products and disinfection. Indeed, the more natural multiple-barrier treatment systems passed down through history are often not recognized. However, good design and management of the entire treatment train will ensure that the design response to regulatory requirements is more likely to be successful.

There are many benefits to rainwater harvesting and some issues to manage. Michelle and Rob have provided the guidance and tools to discover the benefits and clarify the issues leading to good design. We should be mindful that rainwater harvesting is an ancient philosophy that now includes modern technology to help us make the world a better place. It is not a competing philosophy with mains water management approaches, rather it adds to the value and resilience of centralized approaches.

Dr. Coombes has spent more than 30 years dedicated to the development of systems understanding of the urban, rural and natural water cycles with a view to finding optimum solutions for the sustainable use of ecosystem services, provision of infrastructure and urban planning. Learn more at: urbanwatercyclesolutions.com

How to Use this Book

In this book, we've shared the essentials of our design process from start to finish for a rainwater harvesting (RWH) system for a home.

Note that our bias is always to consider a RWH system as part of a much larger strategy for security, resilience, and sustainability, and as such we present not only the nuts and bolts for calculating roof, tank, and conveyance sizing, but also the background strategy and system planning that we think should go into any good design.

The recommendations we make here regarding the design of a RWH system are based on a few things. First, there's our personal experiences combined with the experiences of the many other professionals and installers with whom we have collaborated over the years and whom we trust. In addition, this book reflects decades of research and extensive monitoring done by independently funded researchers, particularly in Australia, where over three million people drink rainwater everyday. The vast majority of these RWH systems do not include end-of-line disinfection components (such as an ultraviolet sterilizer), and the health of Australians reliant on rainwater is the same or better than that of those relying on water supplied by the municipal water system.

A *well-designed* and *sensibly maintained* home-scale, roof-harvested rainwater system requires thoughtful planning for each component (roof, gutters, downpipes, storage tank, pumps, plumbing, etc.). Water quality improves as a droplet of water moves through the RWH system, in part because of a naturally occurring set of treatment processes going on in the rainwater tank. Good design and sensible maintenance practices that contribute to, as well as support, this inherent RWH system *treatment train* will result in water that is of acceptable quality for drinking and other domestic indoor and outdoor uses

without the need for end-of-line disinfection and sterilization.

The focus of this book is therefore to explain what a well-designed system is (one that results in a RWH treatment train) and how to provide sensible maintenance for that system.

However, no matter where in the world we are designing a system, we always have to start by looking at the particular regulatory context of that location, and our designs may shift and change to reflect the requirements, code, and/or standards prescribed by the law there. Therefore, as much as we'd love to be able to say that this book is the only document you'll need to design a RWH system for your home, because of the vastly different regulatory contexts (and conflicting perspectives relating to the safety of rainwater), it is simply not true.

Your local regulations, code, and applicable technical standards may set additional prescriptive requirements for your RWH system; they may even limit permissible uses of rainwater in your home. There may be additional design and installation requirements, as well as operating and monitoring/testing requirements. Your regulator may even mandate that you use end-of-line disinfection (ultraviolet sterilization, ozone sterilization, or chlorine treatment) if you want to use rainwater in your home.

As a homeowner, you need to make informed choices about the correct design for your particular situation. End-of-line disinfection might be warranted — or it might not. This book aims give you all that you need to make informed choices.

Essential Rainwater Harvesting: A Guide to Home-Scale System Design is structured as follows:

Chapter 1. Introduction: An overview of how your RWH system can fit within the larger context of full-property water security and over-all sustainability. In addition, we provide some guidance on how to navigate your local regulatory context.

Chapter 2. Fundamentals: An overview of the fundamental concepts in the design process including water quality, supply, demand, and managing the risk of low rainfall.

Chapter 3. Feasibility: How to evaluate the feasibility of your project and select the size(s) of your roof and storage to best meet your needs for the least possible cost. Here we provide you with the step-by-step instructions to build and use your own RWH system calculation tool, using any spreadsheet software.

Chapter 4. Anatomy of a Rainwater System: How to best select and plan the spatial layout of the individual components on your site so they work together as a whole and the system functions as intended.

Chapters 5. Collection and Pre-Filtration; Chapter 6. Storage; and Chapter 7. Pumps: The nuts and bolts details and considerations for the individual components themselves, including selecting and sizing collection equipment and materials (roof, gutters, downspouts, pre-filtration, etc.), storage considerations (tank and associated fittings), and the basics of how to best select a pump, if one is required.

Chapters 8. Assuring End-Use Water Quality: This final chapter provides a summary of the individual upfront and ongoing practices discussed throughout the book that ultimately result in good design, sensible maintenance, and a RWH system that acts as a treatment train in and of itself. We also provide some recommendations for end-of-line filtration, and, in case you should require disinfection, we've included an example of a disinfection treatment using ultraviolet sterilization.

At the end of the book, you'll find a comprehensive Resources and References section. It contains other recommended books, a listing of applicable North American codes and standards related to RWH systems, our favorite online resources (websites, blogs, and videos) as well as a publications section that contains a relatively small compilation of some of the relevant research as well as references to other useful publications on rainwater harvesting systems.

We emphasize that it is ultimately you (or the homeowner) who is responsible for assuring that a RWH system delivers water quality that is appropriate for its end-use. With this book, we hope to help you make informed choices to design not only a RWH system, but a *good* RWH system — one that meets your needs in your particular context and delivers appropriate water quality, now and into the future.

Website Resources and the Downloadable Spreadsheet Tool

This book is standalone, in that everything you need to build your own spreadsheet calculation tool and perform RWH calculations is included. Follow our step-by-step instructions and in no time you'll be performing storage optimization calculations.

We also have a great inventory of additional resources, spreadsheet templates, RWH videos, and more optimization and sizing examples as well as links to relevant research and publications that we've pulled together and made easily available on our website: www.essentialrwh.com.

In addition, if you want to save time with your feasibility calculations and system sizing, you can purchase the Essential Rainwater Harvesting Tool and hit the ground running. This pre-built spreadsheet tool contains the formatted tables and all of the formulas and logic presented here, plus a few more advanced features pre-programmed and ready-to-go.

Learn more about the Essential Rainwater Harvesting Tool, get our free downloads and additional video tutorials, and access tons of great resources related to RWH system design on our website.

Units

The International Standard of Units (SI) is our preferred system of measurement.

With SI units, RWH design calculations are often simple and straightforward, and it's far easier to perform back-of-the napkin calculations, often without the need to even pull out a calculator.

For instance, in SI units: Roof Area (m^2) × Precipitation (mm) = Captured Rainwater (liters)

In US customary units, this same equation becomes: Roof Area (ft^2) × Precipitation (in) × 0.623 = Captured Rainwater (gallons)

Using a scale ruler to create and read a scale map (something we discuss in Chapter 4) is also more straightforward when doing measurements in metric.

Please note that even Standard 63 — which was published by three US organizations — uses SI units as standard, so we encourage you to use SI units when performing RWH calculations. And so, we will present calculations first in metric, followed by US customary units in parentheses.

However, once you've done your calculations and are preparing to select and make your material choices, the reality is that when it comes to commercial components in North America, the specifications are almost always in US customary units (gal, in, ft, etc.). Therefore, if you are in North America, like us, you may want to do your calculations in SI, then convert numbers to US customary units to purchase materials and communicate with vendors. If you need to switch between units, you can use the conversion table found in Appendix A.

If you simply prefer to stick with US customary units throughout, you'll be able to do so, as we've provided all equations and reference tables in both units.

A Note on Significant Figures and Units

All of the calculation tables and examples have been provided in metric units, with the US customary units in parentheses or square brackets. In the examples presented, both systems of measurement are calculated separately and then rounded to appropriate significant digits on a step-by-step basis. Because of this, small rounding discrepancies may result when comparing final results.

Chapter 1

Introduction

WATER IS A VALUABLE RESOURCE — perhaps the most valuable resource we have — and the collection and use of rainfall has been a part of human history for thousands of years. However, in recent decades, the collection and use of rainwater has diminished greatly due to cost reductions in groundwater drilling and the increased prevalence of municipal centralized water systems. Despite the benefits that have come with these developments, the increased ease of access has also facilitated poor design (or really, a complete lack of design), which has subsequently led to an incredibly wasteful use of both water and water-energy in our modern-day homes and cities.

In developed countries, nearly all communities treat water with indifference, as an infinite resource and/or as a liability. We shed water from our roofs and direct it straight to the storm sewer (leading to floods and sewer overflows), then we turn on the sprinkler to water our lawn. We don't ever consider the energy cost and implications of the water that flows freely from the faucets. We drain groundwater aquifers; we discard nutrient-rich water (perfect for feeding plants and biology) directly into the sewer system; and, perhaps most telling of all, we defecate into water that has been processed or cleaned to drinking standards before flushing it away.

Added to the above, research into water affordability (Mack and Wrase, 2017) indicates that rising municipal supply water rates (attributed to aging infrastructure, water quality, sanitation, and climate change, among other things) could mean that in the next five years the number of US households who find municipal water utility bills unaffordable could triple — to

more than 35%. Nevermind that the cost of replacing aging municipal water infrastructure in the US alone is estimated to be over $1 trillion dollars in the next 20 years (AWWA, 2012).

As current water-supply infrastructure continues to age, glaciers melt, and groundwater aquifers diminish, governments, municipalities, and individuals are starting to realize that capturing and storing rainwater is critical to sustainable, economic, and resilient human habitat.

If we wish to create a resilient future, changing our relationship with water is one of the most important things we can do as individuals and as a community, and it starts right outside our back door.

Water Supply, Security, and Sustainability

Are you trying to provide domestic water to your home and see rainwater as your most cost-effective option? Are you concerned about the resilience of your existing water supply and looking for a backup system (for instance, lack of trust in your municipal water system, or perhaps your groundwater well is dying)? Do you value sustainability and see rainwater harvesting as great way to reduce your resource and energy use? Or perhaps your local municipality has made it illegal to use treated municipal water for non-essential needs such as irrigation, and you require a RWH system to water your garden.

Essential Rainwater Harvesting: A Guide to Home-Scale System Design covers all aspects of RWH system design for your home, whether your goal is water supply, water security, or environmental sustainability. We've distilled years of experience and independent research

1

into a step-by-step approach that includes design thinking, goal setting, system planning, site assessment, calculations, and material selection and sizing for roofs, gutters, downspouts, storage tanks, filtration, and pumps — with special considerations for cold climates.

However, in our consultancy practice, our clients are often looking for more than a simple rooftop harvesting system. They want homes and homesteads that leverage and interact with the environment, producing their own energy and food, harvesting and storing water, cycling nutrients, and restoring the surrounding ecosystems.

Although this rainwater harvesting book is focused on the essentials of designing a rooftop rainwater harvesting system for a house, the upfront consideration of how rainwater capture and storage fits into the broader water and resiliency planning for a property is a crucial first step in the design process. As such, here we present a few brief considerations for full-property water security, resilience, and overall sustainability.

Resilient Systems and Properties

> Where sustainability aims to put the world back into balance, resilience looks for ways to manage an imbalanced world.
> — Andrew Zolli, co-author of
> *Resilience: Why Things Bounce Back.*

A system is a set of interacting or interdependent component parts forming a complex/intricate whole. The components that make up your rainwater supply are a system. The elements that are put together to provide for your shelter, water, waste, and food needs can be thought of as a system, e.g. our homes are systems. Our neighborhoods and our cities are also systems. Everywhere you look, systems are nested within systems, and it's really just a question of where you draw the boundary.

Resilience is the capacity for a system to adapt (and, we would argue, to continue to thrive) in the face of change or disruption. It's an excellent complement to sustainability, and arguably you can't have resilience without sustainability. However, we like to present and think about systems in terms of their resilience because, fundamentally, for many of our clients and students, their primary motivation for taking action is to increase their personal resilience.

To design a system that is resilient, the design must include redundancy and be efficient, productive, appropriate, and interconnected.

To expand on these resiliency characteristics, in Table 1.1 we present design choices and examples for each characteristic for a resilient property and contrast these against design choices for most modest modern-day homes (i.e. a fragile property).

Resilience is the outcome that results when a system includes redundancy and is:

• Efficient

• Productive

• Appropriate

• Interconnected

Table 1.1: A resilient property vs most modern day design.

	Resilient Property	Modern-Day Design
Efficiency	**Focus is on maximizing efficiency.** • Reduction and resource efficiency is the first priority. • Designs are enduring, repairable, solid state, and low tech, where possible.	**Unlimited resource mentality.** • No consideration of quantity of resources used. • Energy and fossil fuel used to make up for design shortcomings. • Design is low quality; elements within the system are disposable, with planned obsolescence.
Productiveness	**The home and occupants collect resources and produce abundantly.** • Home and/or occupants are producers of some or all of their energy, water, and food needs. • Ecological services and products are recognized, valued, and encouraged.	**The home and occupants are merely consumers.** • Constant external inputs required for all needs, including energy, water, and food. • No consideration of ecological yields.
Appropri-ateness	**Energy and water appropriate for end-use.*** • Energy density appropriate to end-use. • Water quality appropriate to end-use. • Gravity used, where possible.	**Energy and water not always appropriate for end-use.*** • No consideration of energy density or appropriate water quality. • Fossil fuel used to make up for design shortcomings.
Interconnectedness	**Design is cyclical and very connected.** • Waste is recycled. • Everything is used multiple times. • Every element has multiple functions and is supported by other elements. • Feedback influences occupants' behavior resulting in beneficial course correction.	**Design is linear and unconnected.** • Waste is sent away. • Single-use mentality. • Designed obsolescence. • Requires constant external inputs. • Lack of integration. • No feedback, and occupants unaware of the consequences of their actions.
Redundancy	**Redundancy is key.** • Backup/alternative plans in place for heat, power, water, and food. • Storage in place for energy, water, and food. • Long-term thinking.	**No redundancy.** • Critical systems have no backup. • Complete dependence on constant, ongoing external inputs. • Short-term thinking.

*Think of how inappropriate it is to cut butter with a chainsaw. The same idea applies to heating your home with natural gas, or using drinking-quality water to flush a toilet. These are poor matches of end-use with energy density or water quality.

The property illustrated in Figure 1.1 is an example of a resilient property located in a cold climate. Note that you can apply the same resiliency characteristics to each individual sub-system for water, energy, and food.

The Water Systems

Efficient: All fixtures high-efficiency/low-volume; water is reused to reduce irrigation and toilet flushing volumes. Gutters and roof designed to prevent rainwater losses. Water-wise landscaping.

Productive: All water captured (building and landscape) and stored for use. Water employed to grow abundant plant and animal life.

Appropriate: Highest-quality filter located at drinking water tap. Greywater used for toilet flushing and irrigation. Rain used for irrigation.

Interconnected: Wastewater is a resource, and nutrients from wastewater are used to grow plants and soil both in the greenhouse and in the landscape. Monitors in place for water usage, rain forecasts, and storage volumes.

Include Redundancy: Water storage designed for appropriate low-rainfall conditions. Backup plan in place.

The Energy Systems

Efficient: High-insulation walls and windows and careful design of the building envelope. Energy-efficient lighting and appliances.

Productive: House oriented to capture passive solar energy. Careful design and selection of glazing. Attached greenhouse provides supplemental heating. Solar photovoltaics for electricity and solar thermal for hot water.

Appropriate: Passive solar gain as primary heating source, biomass as backup heating, photovoltaic electricity used for high-energy demands, solar thermal used for domestic hot water and ancillary space heat. Irrigation is provided passively (using gravity vs being pumped).

Interconnected: Heat-recovery ventilator used to pre-heat intake air. Warm, stale air from the house cycled to the greenhouse. Excess heat energy from the greywater captured in greenhouse.

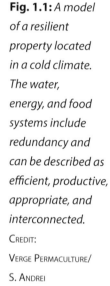

Fig. 1.1: *A model of a resilient property located in a cold climate. The water, energy, and food systems include redundancy and can be described as efficient, productive, appropriate, and interconnected.*
CREDIT:
VERGE PERMACULTURE/
S. ANDREI

Monitoring systems in place for energy production and storage.

Include Redundancy: Grid-tied power for backup electricity. Heating provided by three sources: passive solar, active solar, and biomass.

The Food Systems

Efficient: Local food and seasonal food prioritized. When in abundance, food grown onsite is harvested, preserved, and stored for later use.

Productive: Perennial forest gardening, annual vegetable production, and four-season production in the attached greenhouse. Micro-livestock systems (such as chickens) for eggs and meat. Biodiverse and abundant plant and animal life grown not only for the occupants, but to support surrounding ecosystems.

Appropriate: Food that supplies the occupants is grown with appropriate energy (human-scale vs large manufacturing-scale) and appropriate water (local and captured water vs imported water). Gardens designed for passive irrigation instead of pumped irrigation.

Interconnected: Food scraps, plant, and others wastes are cycled to create compost, soil, and/or to feed micro-livestock.

Include Redundancy: Not purely reliant on industrialized food system. Occupants grow some of their own food and support the local food economy as much as possible.

If your motivation is resilience, you'll want to keep these characteristics in mind as you are designing your RWH system, but also particularly as you consider your RWH system in the context of full-property water security and overall property design.

Order of Design Priorities

For a property to be truly sustainable and resilient, its occupants must have a safe and dependable supply of water, along with protection from the dangers of drought, flooding, and erosion. Recognizing the management of water — a source of life, and source of risk — as central to the success of an environmentally integrated system, it is always the first consideration when we design any property.

There are many elements to consider related to water when in the early design stages: potential water sources (municipal water, groundwater, surface water, rainwater, etc.), watershed management, landscape water retention/diversion/distribution, land-shaping (ponds, swales, collection, drains), infiltration, overflow management, storm water, water reuse, septic, and more.

In practice, we most often see homeowners purchase property, build a house, then try and figure out how to manage and supply water to that structure. The correct order of design priorities (water first, infrastructure next) is completely missed in most modern-day design. To most successfully achieve sustainability or resilience, water planning must be a priority in the *initial* design stages. An understanding of how volumes can be minimized/optimized, where water will come from, where it might be stored, and where it will go should be considered before even starting out on the design and placement of any building.

Water-Harvesting Earthworks

If the scope of your project is larger than a simple rooftop RWH system, and you (1) want your home integrated within a lush, biodiverse environment that restores fertile soil, sustains plants, wildlife, and humans, and you'd (2) like your property to be drought-proof and/or minimize the risk of fire, then you'll absolutely want to consider how your rooftop RWH system fits in as part of a much larger integrated water design.

One of the ways we do this in our consultancy practice is by thinking of the landscape and soils surrounding the home as the primary water *storage* system. We can enhance and improve water infiltration and storage with careful consideration of property elevations and any appropriate water-harvesting earthworks such as rain gardens, swales, diversion drains, ponds, etc.

Our favorite resources for planning and designing landscape water-harvesting earthworks are included in the Resources section at the back of this book.

Improving Soil Water-Holding Capacity

Whether rain falls directly on the ground, or is directed to landscape from tank overflows or water reuse strategies, if the ground surrounding your home is hard and compacted, the water will not infiltrate and will be virtually ineffective. Therefore, when it comes to landscape hydration for supporting biology, gardens, food production, or ecosystem services in general, your best bet is to make sure that the soil on the property is healthy and has high water-holding capacity.

How to do this? Well, soil water-holding capacity increases significantly with even a slight increase in soil carbon content. A 1% increase in soil carbon on your property will store an additional 168,000 liters/hectare (17,960 US gal/ acre) (Jones, 2010). This also makes soil storage the most cost-effective way to increase your overall property water stores. Consider that to store this same volume of water in tanks would cost you anywhere from \$30,000–\$70,000 in infrastructure.

Because carbon content increases when you support soil biology and soil health, there are three primary practices for increasing the water-holding capacity of your land:

1. Keep the soil covered (no bare soil) with living plants, year-around if possible.
2. Maximize diversity in crops and plant species.
3. Avoid the use of synthetic fertilizers, fungicides, insecticides, and herbicides.

Best of all, adhering to these practices not only increases the water-holding capacity of your landscape, it also builds upon and enhances your natural capital and increases your overall sustainability and resilience through increased biodiversity, improved ability to grow nutrient-dense food, and reduced atmospheric carbon content, turning your "footprint" into a positive one.

Learn More

Again, we advise you to remember the importance and benefits of considering your rooftop RWH system as simply a small part of a much larger strategy for full-property water security, sustainability, and resilience. See this as an opportunity to create an integrated natural ecosystem that buffers you against the impacts of disaster while it increases the long-term economic and environmental value of your property.

In addition, changing our relationship with water has enormous implications and a large positive environmental impact when you consider the global crises we are facing with respect to biodiversity loss, food insecurity, and climate change. But going forward, we are going to narrow our focus in this book and keep it to the essentials of rooftop RWH system design.

Make sure you check out the other titles in the *Sustainable Building Essentials* series, and for more information on full-property design, including our favorite books, recommended courses, resources, and blogs on related subjects such as permaculture design, sustainability, soil and soil health, see the Resources and References section at the back of this book.

Design Scenarios

Now that you understand how the RWH system fits (or may fit) within the larger context of water design for your property, it's time to narrow down your RWH design scenario.

Are you in a remote off-grid location? Are you in an existing house in the city? Are you building a new building or retrofitting an existing building? What are your water supply options?

What about your climate? Are you in a cold climate with ground frost in the winter? Is it an arid climate? Are your rainfall patterns uniform throughout the year or does rain come in certain seasons?

Depending on the answers to the above, your RWH system may be simply a small piece of a household water-supply plan, or it may be designed to supply all of your water needs at all times.

There are many permutations of a RWH system, but the three most common household scenarios we see with our clients are:

1. Supplemental supply: rainwater as the secondary system, with a primary system in place.
2. Primary supply: rainwater as the primary system, with secondary system in place.
3. Off-grid supply: rainwater as the primary system, with no secondary system.

In the above scenarios, the primary or secondary system could be a municipal water supply, a groundwater well, gravity-fed or pumped system from surface water on the property, or even trucked-in water.

In Scenario 1, the RWH system is there for specific or supplemental water supply, such as irrigation. Sized usually to only meet specific loads (like toilet flushing), its intended use is either as supplementary supply or as an emergency system, or as both.

In Scenario 2, the RWH system could be designed to meet most (if not all) of the demand over the course of the year, and the secondary system is in place to shave off the peak demands, as well as kicking in in the event of drought or low rainfall. This can be a cost-effective strategy for supplying a majority of your needs with rainwater, especially when a municipal or groundwater supply is already in place.

In Scenario 3, the RWH system must supply all of the demand, all year round. Here, there is a far lower risk tolerance for a zero rain tank balance than in scenario 1 or 2, and this must be factored into the design of the system.

Each of these most-common RWH design scenarios has different functionality and different risk considerations. In the next chapter we'll present typical design considerations and approaches for each. And even if your particular design and usage scenario is not exactly as described above, you should easily be able to apply the presented design thinking and strategizing to your particular needs.

Regulations, Codes, and Standards

Before starting on any RWH project, you must review and understand the local regulations and legal minimum technical requirements. Unfortunately, this is not always straightforward. Depending on where you are in the world, the legal framework for RWH can be clear-cut, complex, contradictory, or — more often than not — quite ambiguous.

Added to this is the fact that significant legislative reform is happening and/or anticipated, particularly in North America, given the increased awareness around water issues and impending shortages.

When investigating and attempting to navigate your current local legal requirements, it is helpful to understand the typical structure of regulations, codes, and standards, and how these relate to one another. First off, there are usually (but not always) regulations such as local laws and by-laws that regulate the rights and allowable uses for captured rainwater. For example, provincial, state, or municipal regulations might stipulate the following:

- Who owns the captured rainwater (the land-owner? the government?).
- Conditions and requirements pertaining to rainwater capture and use, if any.
- The minimum applicable technical requirements for RWH systems.

As for that last bullet pertaining to the technical requirements: regulators often point to an existing code (or possibly multiple codes) written and maintained by third parties. For instance, the Province of Alberta has adopted The National Plumbing Code which is written, maintained, and issued by the National Research Council Canada.

Be careful, however, because many local jurisdictions adopt a code and then apply modifications, variations, and interpretations that take precedence over the base code itself. In addition, there may also be several different regulations that point to different codes pertaining to different aspects of a RWH system. For instance, there could be different regulations/codes related to plumbing, electrical, tank, water quality, safety, and health. If you are lucky, your jurisdiction will have published some additional guidelines or documentation to help you navigate the requirements for RWH systems.

Once you get your hands on the appropriate regulations, codes, guidelines, and any other publications for your jurisdiction, you'll discover that these documents often specify minimum requirements and attributes, but not much more. This is where standards can come in. A standard is a document, usually written and/or recognized by a technical society, that provides increased information and/or requirements for the design, materials selection, methods, etc. In 2013, the American Rainwater Catchment Systems Association (ARCSA), the American

Society of Plumbing Engineers (ASPE), and the American National Standards Institute (ANSI) jointly developed *ARCSA/ASPE/ANSI 63: Rainwater Catchment Systems* (also known as *Standard 63*). In mid 2018, the Canadian Standards Association (CSA) and the International Code Council (ICC) published CSA/ICC B805 Rainwater Harvesting systems. It's important to know that a standard is only legally binding if it has been enacted as such by your local legislation. Otherwise, the standard is purely advisory. It's sometimes worth looking at who was on a standard committee to understand any inherent bias that the standard may possess.

So, when getting started, go straight to the regulatory framework for your jurisdiction, and follow the rabbit-hole from there. Find out if rainwater capture is even legal, if there are restrictions on certain uses, and if there are requirements, regulations, and codes that must be satisfied. Make sure you understand any local variances to the code. Also worth noting is that even though you may run into a regulatory road-block, there are often alternative compliance pathways — be sure to ask your regulator about this possibility.

In addition to rainwater-specific legal requirements, you may also need to consider general developmental approval or planning permission that may be required by your jurisdiction. For instance, installing a tank on your property might require a permit and/or inspection to ensure that the applicable codes and standards have been met and/or that the placement of your tank meets minimum property boundary distances.

Because this isn't always straightforward, a good starting place is to seek out your local rainwater harvesting advocacy group — who, hopefully, have already done some of this research for you. We've also included a listing of the most commonly referred-to codes and standards relating to the installation of RWH systems in North America in the Resources section of this book.

We reiterate that the important thing to know is that your local laws may supersede all of, or parts of, common rainwater harvesting standards or practices, and you can't use information from any book or document on the technical design of a RWH system without an understanding the larger regulatory context that you find yourself in. Also, standards and codes start from the baseline that you've already decided to build a system, and therefore they contain no useful information on establishing feasibility, optimizing performance, or the design process as a whole. That's one (of many) reasons you'll find this book particularly useful.

Chapter 2

Fundamentals

IN THIS CHAPTER, we present the preliminary considerations and some of the fundamentals you'll need to understand before jumping into calculating the system sizing and other requirements. This will help you determine what benefits you might expect depending on your system set-up and how to approach the design process as a whole.

Water Quality

There are many things that will degrade — and some things that will improve — the quality of the rain-harvested water that comes out of your tap.

Some things that may degrade your water quality are:

- Contamination by biological material, such as leaves and decay left on the roof and in gutters, as well as feces from birds and other small animals deposited on your catchment surface or directly into your tank.
- Pollutants from the atmosphere. This is particularly a concern if you live near an industrial area or in a rural area where agricultural sprays are commonly used.
- Compounds leached from RWH materials. Your roof, your gutters, your tank, and your piping (or the coatings in/on them) may leach contaminants into your water, with some materials leaching more than others, particularly when exposed to UV.
- Compounds leached from the materials of your indoor plumbing fixtures. Although perhaps not technically part of the RWH system itself, know that even your indoor plumbing (indoor copper or PEX piping, etc.) and your

hot water tank may leach compounds into the water that exits your tap.

Regardless of these potential contamination sources and risks, a well-designed and sensibly maintained home-scale RWH system has been shown to act as a treatment train and deliver water of a quality suitable for many end uses, including drinking (Coombes, 2016; enHealth, 2012; Morrow, 2012; Evans, et al., 2009; Morrow, et al., 2007).

What does it mean to be well designed and sensibly maintained? Answering that is, in part, the purpose of this book. But to summarize, and give you a sneak peek, as the designer and the owner of the RWH system, your job is to:

- Minimize the contamination sources and vectors through your upfront material and design choices (such as the layout of components and the inclusion of components like screens).
- Design for easy maintenance, including consideration of maintenance during layout and component choices.
- Support the inherent and naturally occurring treatment processes going on in your tank. This includes:
  - Design and operation that minimizes contamination inflows into your rain tank.
  - Design and operation that encourages sedimentation (and the resulting removal of heavy metals from the water column).
  - Design and operation that supports the actions of the functioning ecosystems within the tank (these ecosystems provide vital water-cleaning services).

○ Design that ensures that the sedimentation layer is not disturbed when water is drawn from the tank.

○ Routine maintenance on an ongoing basis.

If you do all of the above, and follow the good design and sensible maintenance practices outlined in this book, your harvested rainwater will be clear, will have little taste or smell, and will be of good quality, without the need for end-point ultraviolet (UV) disinfection, ozone disinfection, or chlorine sterilization.

Thinking About Demand

Demand is the first side of the equation when it comes to rainwater harvesting. Demand can be thought of as simply: *What am I planning on using this water for?* Examples of typical uses for rainwater include: domestic supply where high quality is required (drinking); domestic supply where mid- or lower-quality water is acceptable (toilet flushing, washing machines, lavatory faucets, etc.); and irrigation. Your local laws may, however, disallow certain uses for rainwater

(such as drinking or showering), which will narrow your early design options.

More specifically, to determine demand for your intended uses, you'll need to determine:

• the volume of water needed
• the quality of water needed
• seasonality (the volumes required over time)

Domestic water use typically stays constant in a household over the course of a year, as long as the number of occupants stays relatively constant. For instance, a family can be expected to use approximately the same amount of household water (showering, cooking, toilet flushing) each month over the course of the year. An exception to this would be a vacation home or other seasonal dwelling.

Water volumes can also vary quite significantly based on the types of fixtures (i.e. high-efficiency vs standard). A common example of this is an ultra low-flush toilet vs a standard toilet.

Irrigation volumes will depend on the land size, aridity, and soil carbon levels, as well as plant species; these demands are usually seasonal. (And remember that healthy soil will require less water.)

When heading down the rainwater harvesting path, you'll quickly realize that the more water you need or want to supply (i.e. the larger your demand) usually means a larger system and more cost. So reducing your demand — or at least critically evaluating what your demand actually needs to be — is the very best way to minimize the cost of your RWH system.

How to calculate/estimate demand volumes is covered in detail in Chapter 3, but first let's consider some major factors affecting water usage volumes.

Behavior and Feedback

In our modern-day lives and homes, most of the services we depend on (water, electricity,

Potable Rainwater

Your private, home-scale RWH system may — or may not — be regulated by your local authorities. If it is regulated, your system will likely fall into one of two classifications: *potable* or *non-potable.*

Potable RWH systems have stringent regulatory requirements in terms of the materials used for the roof, gutters, piping, fittings, valves, tanks, etc. Sometimes this means that you must select, purchase, and use materials that come with a manufacturer's third-party-certified potable rating.

In addition, your regulator may require you to provide end-point disinfection or sterilization — regardless of the actual water quality your RWH system delivers. Regulatory requirements for filtration and disinfection are covered in more detail in Chapter 8.

waste, natural gas, propane, heating oil, etc.) are provided by centralized systems. These systems satisfy our needs by generating power, pumping water, or treating waste in facilities far away from our homes. Although we may enjoy all this convenience, the ramifications of participation in centralized systems are mostly removed from view.

Feedback is the process in which the effect of an action is returned to the doer. We consider feedback to be a primary and fundamental principle that must be present in good design because it has drastic and positive implications on behavior.

For example, individuals living with solar photovoltaic systems reduce their power consumption; solar thermal users change showering and bathing patterns; greywater and septic system users drastically reduce the amount of toxic chemicals used in the home and discarded down the drain; and — you guessed it — rainwater system users drastically reduce the amount of water they consume, simply through changes in behavior.

We've found in practice that the actual volumes of domestic water used in households using rainwater in prosperous parts of the world is far less than the typical or average household volumes reported in engineering standards and technical design tables.

Don't forget that there are also circumstances that will increase domestic water usage *above* an anticipated average — circumstances that you have no control over as a designer. Examples include extra guests for an extended period, social gatherings, and hot-weather periods.

Reducing Demand Through Design

Our landscapes and our homes can also be set up to use less water right from the onset. The reality is that our conventional household fixtures are huge water-guzzlers.

Landscape and irrigation volume requirements are substantially reduced by employing the strategies discussed in Chapter 1 for full-property water security. To reduce the water used in the home, you can change to or select low-water-use fixtures. The amount of water used by a low-flush toilet can be one-third that of a standard toilet. The same goes for regular washing machines vs high-efficiency ones. These types of water efficiencies are significant when calculating expected usage volumes.

Off-grid rainwater-supply scenarios should absolutely employ numerous conservation strategies, including using low-volume fixtures throughout the home and perhaps even a composting toilet. In addition, water that is still relatively clean after its original use (such as after hand washing), should be directed to a secondary use (such as irrigation).

The term *greywater* is used to describe water that is redirected from lavatory faucets, showers, bathtubs, and washing machines to the landscape for irrigation. A greywater strategy can substantially reduce the demand loads for a RWH system. One of the most innovative thinkers we have run across in the area of water conservation and reuse is Michael Reynolds, pioneer of the Earthship concept (see Resources for more information). Reynolds has thought very deeply about how to harvest, capture, and use water multiple times in a building to get the most out of every drop. His systems combine rainwater harvesting, greywater collection and processing, blackwater utilization, and food production. Mapping the flow of water in an Earthship is a fascinating exercise that will open your eyes to conservation and reuse in water design. See also Figure 1.1 in Chapter 1, where many of these same ideas are applied to the design of a resilient and sustainable home and landscape.

Despite what has been shown to be possible, there are many places and municipalities that do not permit water reuse. Some only permit certain types of water reuse, or have requirements for treatment (chlorination, for instance) prior to reuse. Readers are advised to always inquire about and follow location-specific laws and regulations pertaining to greywater and water reuse.

See the Resources section for more information on using and designing greywater systems for your home.

Thinking About Supply

In order to estimate the amount of rain that you expect to collect, you'll first need to get your hands on an appropriate rainfall dataset for your location. A rainfall dataset is the set of numbers that describes your average rainfall volumes and associated patterns, expressed over the course of one calendar year.

Note that the rainfall on the first day of spring last year may look very different from the rainfall on the first day of spring the year before. To account for this annual variability but still represent an accurate overall pattern, multiple years of past rainfall data are usually averaged to create an average, calendar-year rainfall dataset. The number of years used to calculate this average is called the *duration.*

Also, whereas in real life we experience rainfall instantaneously (as it is happening), on paper we express rainfall as a depth over some set amount of time. For instance, you'll see rainfall defined as mm per second, mm per minute, mm per hour, mm per day, or mm per month (or inches, of course). The size of the time interval (second, hour, month, etc.) is the *time step.*

Here are some examples of possible time steps and durations and the subsequent number of data points in the final calendar-year rainfall dataset.

- Monthly time step based on 1-year duration (12 data points: Jan–Dec)
- Monthly time step based on 30-year duration (12 data points: Jan–Dec)
- Daily time step based on 15-year duration (365 data points: Jan 1–Dec 31)
- Hourly time step based on 15-year duration (8,760 data points: Jan 1 01:00–Dec 31 24:00)
- Five-minute time step based on 30-year duration (105,120 data points: Jan 1 00:05–Dec 31 24:00)

It's intuitive that the last dataset in the list above would provide far more accurate supply estimations than the first one. In fact, research into the simulation of RWH systems has shown that performance modeling is critically dependent on both the time step chosen and the duration (Lucas, et al., 2006). However, there's a pragmatic trade-off, particularly for the average home-scale RWH system designer. When seeking out rainfall data and performing calculations, selecting too small of a time step will result in data management and computational problems. Although a five-minute time step may provide high accuracy, it would require 288 data points per day, or over one hundred thousand subsequent computations to model a full year of supply and demand. Also, data over small time steps is much harder to get your hands on. On the other hand, selecting too large of a time step results in an unwanted averaging effect. The low granularity of the data increases the possibility for erroneous results and inadequate design.

Despite the inherent gross simplifications, we usually recommend that you seek out rainfall averages in mm/month (in/month) over the course of 30 years, in part because this historical dataset is often made readily available by most governments. We'll show you where to find this dataset, and discuss limitations of using this level of data granularity in Chapter 3.

Rainfall Pattern, Catchment, and Storage

There's four important concepts to understand when it comes to supply: *rainfall pattern, catchment, storage* and perhaps most importantly, the *interplay* between rainfall pattern, catchment, and storage.

Rainfall Pattern

Your rainfall pattern is described by both your total average annual rainfall, mm/yr (gal/yr) and the percentage distribution of rain on a calendar-year basis. It is entirely dependent on where you live and is something that you have no control over. To give you a better understanding of how rainfall patterns differ from location to location, Figure 2.1 shows the monthly rain distribution for four different cities. The total average annual rainfall for each city is stated in the legend beside the city name.

You may live somewhere that receives a fairly steady amount of rain every month of the year, and total amount of rainfall is moderate to high (like Worchester, Massachusetts). Or perhaps your climate provides a moderate-to-high amount of rain that is unevenly distributed and

most of it comes in the winter months (like Vancouver, British Columbia). If you live somewhere like Flagstaff, Arizona, you receive only a small amount of rain on an annual basis, and it tends to come in early winter and late summer. From a design perspective, places like Amman, Jordan, have the most challenging rainfall patterns, as they receive virtually no rain for half of the year, and then they have several very large rain events during the other half of the year.

Catchment Area

The catchment area is the total area from which you can harvest your rainwater. In home-scale rainwater design, your catchment area is basically the footprint of your roof, as it's impractical to attempt to harvest rain from other surfaces such as your driveway (although something like this might be done in a commercial system). Note also that harvesting water from surfaces other than your roof is likely to result in a

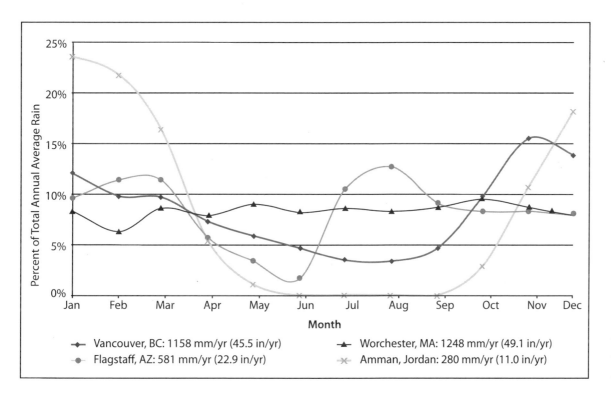

Fig. 2.1: *Percentage of total annual rainfall on a month-by-month basis for four cities. The total annual rainfall for each city is shown in the legend.*

dramatic reduction in water quality. When we say: *Increase the size of your catchment,* what you need to do is assess if you can increase the footprint of your roof by changing the shape of your building, or by adding other roofs, such as a shed, a garage, or perhaps even a neighbor's roof to your system (if it's practical to do so).

Storage

The primary role of your storage is to smooth out supply and demand. Think of it this way: on a second-by-second basis, you'll potentially have rain coming in (supply) and rainwater going out (demand). Sometimes you may have a positive net supply (more supply than demand) and other times a negative net supply (less supply than demand). Your storage is used to carry-over the net surplus or make up the difference in the event of a water deficit.

The tank is typically one of the most expensive elements in a system, especially in larger RWH systems — even more so in cold climates because tanks must often be buried below the frost line for freeze protection. Therefore, when selecting a tank, you'll almost always want the smallest storage volume that will meet your demands over the course of a year.

Rainfall Pattern, Catchment, and Storage Interplay

If the owners of a home in each of the four cities presented in Figure 2.1 were designing RWH systems to supply approximately equivalent domestic demands, here is how their systems would compare:

- Worchester, MA: Smallest storage.
- Vancouver, BC: For the same catchment area as the Worchester home, moderately larger storage.
- Flagstaff, AZ: For the same catchment as the Worchester home, significantly larger storage.

- Amman, Jordan: Would require at least double the catchment area and an enormous storage capacity compared to all of the others.

Space and Cost

Your ability to increase supply by increasing catchment, or your ability to "smooth" out your rainfall pattern by increasing storage might be space-constrained from the onset. For example, if you are adding RWH to an existing urban home on a small lot, your roof size is fixed, and you may be limited by the size of tank that can actually fit onto your property.

Alternatively, if you are planning a new home, you'll likely want to use your RWH system design to inform your building shape and the resultant roof footprint; this initial stage is the easiest time to add an additional secondary roof to your system to meet your demands.

If you are planning on year-round rainwater harvesting where frost protection is required, you'll definitely need to think about where you are going to put your tank. You'll need to protect your tank from freezing by burying it, integrating it into a basement, or by some other means. Regardless, it requires substantial upfront space-planning.

We often don't appreciate just how much water we and our conventional homes consume. When we attempt to meet our water demands on a rainwater budget, we sometimes discover that the required catchment area would be absurd and/or the size of tank required would be completely unfeasible from a space perspective, or from a cost perspective, or both.

Managing Your Risk
Using Averages in Design

When it comes to predicting future rainfall, we've already suggested that you seek out monthly rainfall averages in mm/month (in/month) over the course of 30 years, and use those numbers as the starting point for your design calculations.

But before proceeding blindly with averages, it's well worth looking a little closer at your dataset to get a sense for the amount of actual variation or spread that might be considered normal for your particular climate.

Table 2.1 shows the monthly rainfall averages in mm (inches) for Calgary, Alberta, for the month of May for the ten-year period 1990–2000. If you were using this ten-year duration for your dataset, you'd use 50 mm (2 in) as your average rainfall volume for the month of May in your preliminary design calculations.

Note however that, within this ten-year dataset, there were four years when the rainfall was less than 30% of the monthly average, and two years where the rainfall was half. That's quite a substantial deviation, especially given that this is a relatively small dataset. A larger dataset would likely show more spread and a larger variation from the mean. Depending on how robust you want/need your RWH system to be, the year-to-year variation might be a major consideration in your design. We'll discuss how your design can compensate for this rainfall variability in Chapter 3.

Therefore, we highly recommend that, in addition to looking at the average multi-year precipitation number, you always dig a little deeper into the data to get a sense of what size of variation is historically normal. It might not be unreasonable at all to even look at the daily rainfall data, especially if you have access to that data.

Extreme Rainfall

Water forces are powerful, and mishandled surge volumes can compromise tank and building foundations, or cause serious erosion on your site. No matter which RWH design scenario you employ (supplemental, primary, or off-grid),

Table 2.1

Year	Average Precipitation in May, mm [in]
1990	91 [3.6]
1991	21 [0.8]
1992	34 [1.3]
1993	62 [2.4]
1994	62 [2.4]
1995	60 [2.4]
1996	26 [1.0]
1997	32 [1.3]
1998	86 [3.4]
1999	47 [1.9]
2000	29 [1.1]
Average	50 [2.0]

Table 2.1:
Monthly rainfall average in mm [inches] for Calgary, Alberta for the month of May, 1990–2000.

you want to ensure that you design your gutters, conveyance, and tank overflow appropriately to minimize the potential of damage to infrastructure and landscape.

Gutters, conveyance piping, pre-filtration, and tank overflows are sized based on the maximum rainfall intensity that you might reasonably expect. These topics are covered in Chapter 5.

If you are sending your tank overflows to the landscape, such as rain gardens or swales, you'll also want to consider the likely available retention capacity in your tank prior to a significant rain event combined with the estimated infiltration capacity of your soils. However, designing for landscape infiltration is outside of the scope of this book. See the Resources section for more information.

Drought

We've found that historical drought data is much harder to get your hands on than rainfall data. Drought is often not even included as a consideration in many RWH design resources. We assume that's because it's not as technically relevant from a system-sizing perspective (i.e.

too little rain won't wreck your infrastructure or cause foundation failure).

However, a failure to supply the water you actually need can be critical, depending on your context and scenario. For example, for primary systems with backup and/or where your rainwater is simply a supplemental system, you may not need to worry too much about the drought scenario. However, if you are designing for a remote off-grid acreage on an island with no groundwater, you'll want to thoroughly consider how to most reasonably (and cost effectively) manage your risk in the event of a long-term supply shortage.

Minimum Rain Limit

Stack all of the above on top of the general concern that the climate is shifting and precipitation patterns are changing, and you can see that simply using averages to design your system might be a bad idea.

On the other hand, there's another pragmatic trade-off between how much you want your design to handle the worst-case drought condition and how much you are willing to pay for it. For example, it might be tempting to state: *I want my system to meet all of my demand needs even if there is a 100-year drought.* The reality, though, is that achieving this goal is likely to be cost prohibitive and may not even be technically possible.

So, as part of the design process, you will have to make a decision about what reduction in rainfall is reasonable for you to design your system to handle. We call this the *minimum rain limit*. And although you might not have any idea what is reasonable to expect at this point, it will be helpful later down the road if you take some upfront time to consider the following:

- Look at the rainfall variation in your dataset. How big is the spread between the average and the low? How often does a low rainfall condition occur? How long did low rainfall conditions last in the past?

- Talk to longtime locals. Farmers and gardeners can be great resources and may even confirm qualitatively and perhaps quantitatively the severity and the lengths of any droughts that have occurred in the past.

Establishing a Minimum Operating Goal

A *minimum operating goal* is a statement about how you want your storage to perform at your minimum rain limit.

Unfortunately, there's not a prescriptive formula for what you should state here. Your context, your goals, and your values, combined with an understanding of historical variations in rainfall data for your climate will ultimately inform your minimum operating goal.

You should ask yourself: What is the consequence of not meeting my demand needs? Is the cost of importing water substantial? Other questions to consider are the other backup supply options available and the costs to drill a well or tie into municipal water.

You'll want to evaluate these things, and at least have an idea of cost and availability of water-supply alternatives before you set a minimum operating goal for your RWH system.

Also important is that you state your minimum operating goal in terms of the storage performance you expect from your system at your minimum rain limit.

Here are some examples of well-stated minimum operating goals:

- I'd like to store enough water to be able to irrigate for two months even if there is no rainfall in that period.

- I want three months of supply available and stored in the tank in the 20-year drought scenario.

- I want to continue to meet my demands even if rainfall is 50% of normal for two years.
- I only want to switch to my backup system if there is a three-month drought.
- I want my system to meet 40% of my demand needs in the 50-year drought scenario.

There's also another perfectly acceptable minimum operating goal: *I want my system to meet my demand needs as long as rainfall is close to the average expected amounts.* What you are really saying is that you are not too concerned about the low rainfall occurrence, likely because your demand needs are supplemental (or non-critical), because you have an existing backup system, or because you are willing to change or adjust demands during different rainfall conditions.

To give you a glimpse of how this all ties together, the minimum rain limit and the minimum operating goal help us to better optimize the design that starts out based on average rainfall conditions. And often — but not always — you'll discover that in order to meet your minimum operating goal, you have to tweak your design.

The good news is that because the design process is iterative (you repeat the calculations using different parameters), you can always start with some ideal resiliency, and then, if the system you come up with is cost prohibitive, you can lower your expectations until you find a design that works within your budget.

Adaptive Strategies

Research has shown that users of RWH systems incorporate the monitoring and awareness of short-to-medium-term weather forecasts into the management and operation of their RWH systems. Therefore, before adding a ton of capacity (and potential cost) to your RWH system to meet your minimum operating goal, it's well worth discussing what we call *adaptive strategies.*

Adaptive strategies are real-time behaviors or backups that can be put in place in the event of a rainwater-supply disruption. Regardless of your RWH supply scenario (supplemental, primary, off-grid), your typical adaptive strategies for managing lack of rainfall or drought would be to:

- Employ water conservation: Reduce shower duration and frequency (changing from a six-minute shower to a three-minute shower is a massive savings in water use); reduce or eliminate irrigation; wash your laundry at a laundromat; wash dishes by hand; shower at the gym or at work, etc.
- Import or purchase water. This could be on a small scale (such as a bottled water supply) or on a large scale (such as bringing in trucked water).
- Switch to a backup system. Of course, this option would only be available if you included a backup system in your design.

For some for households (including many of our clients), suggesting that they shower less in the event of low rainfall is not always a warmly greeted design proposition. However, adjusting actual water usage is absolutely an acceptable risk strategy for a seasonal off-grid cabin or perhaps acreage owners looking to build a DIY system on the smallest possible budget.

The ease or cost of importing water will be entirely context dependent and is a definite factor for how much you decide to rely on adaptive strategies. If you are near a town or major center, hiring someone to deliver water several times a year might be very reasonable and cost effective. If the cost of bringing in water is prohibitive however, you'll certainly be looking to rely more heavily on designing a RWH system that meets your minimum operating goal.

If you are adding a garden irrigation RWH system to an existing home with municipal water supply, you may not be overly concerned about the risk of having no water in your rain tank because you will be able to simply switch to the other system already in place.

The main idea here is that if you are willing/able to rely more heavily on behavior-based strategies, you can likely find a suitable design with a smaller roof area and/or tank as well as reduce the size or even the necessity of a backup system. This could mean a lower upfront capital cost for your water-supply system.

To provide maximum optionality, scalability, and adaptability of your system in the future — no matter what your minimum operating goal is — you'll absolutely want to consider including level monitoring and spare connections

on your tank. Monitoring is key to providing feedback and is therefore an important upfront design element. Spare connections should allow for the flexibility to add more storage (more tanks) or potentially tie-in more catchment (more roof) down the road.

Design Tools and Materials

There are a few basic tools that you'll want to get before getting started.

Pencil, Graph Paper, Clipboard, and Scale Ruler

You'll want a pencil and eraser as well as a pad of ledger-sized graph paper (279 × 432 mm [11 × 17″]) and a clipboard of the same size (letter-sized paper never seems to be large enough for a drawing and notes).

You can easily build a clipboard large enough for the ledger paper by using a ¼″ 11 × 17 medium-density fiberboard (MDF) with a binder clip. It is inexpensive — and indispensable when doing field assessments, scale drawings, site plans, and elevation plans.

Architect's Scale Ruler

An architect's scale ruler is another indispensable tool. It is basically a specialized ruler that makes drawing something to scale, or measuring something on a site plan an absolute cinch. Look for a metric scale ruler if you plan on doing your measurements in metric.

Measuring Tape

Get yourself a 30 m (100 ft) and an 8 m (25 ft) tape measure. You'll very likely use both.

Hand-Held Sight Level

We all know that water flows downhill. However, when you are onsite, it can be surprisingly difficult to estimate elevations and/or distances and evaluate placement to meet

Fig. 2.2: *Architect's scale ruler.* CREDIT: VERGE PERMACULTURE

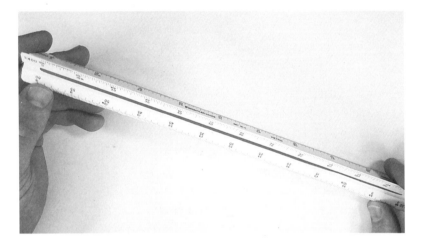

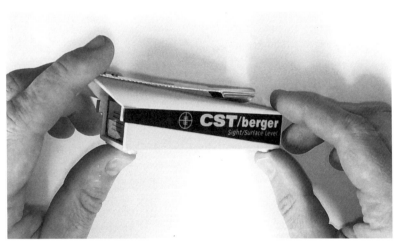

Fig. 2.3: *Hand-held sight level.* CREDIT: VERGE PERMACULTURE

minimum slope requirements, particularly for conveyance piping. This is where a hand-held sight level, combined with a measuring tape, are great tools.

We prefer non-magnified versions, and they typically they cost $30–$60.

Computer-Aided Design Program

In all honesty, all of your planning can be done on paper with a scale ruler and a pencil with an eraser. We still do a lot of designs this way.

However, there are considerable advantages to using computer-assisted design (CAD) software. Once you become familiar with the software you can save quite a bit of time because adjustments and changes to the design are quick and easy and don't require you to erase and redraw lines.

There are two types of drawings that will are most beneficial to you when planning a RWH system: a site plan (view from space), and an elevation plan (view from one side). A piping and instrumentation diagram (P&ID) is a third type of drawing that is sometimes useful as well.

There are many CAD and drawing software programs out there; some are free (like SketchUp), and some are very expensive (like AutoCad). When we want to use CAD, we use Smartdraw, as it is relatively inexpensive and does most everything we need.

Using a Spreadsheet vs a Calculator

Although everything presented in this book could be done by hand with a calculator, it's incredibly impractical to do so given the iterative nature of RWH system design. As such, spreadsheet programs like Microsoft Excel, Mac Numbers, Google Sheets (free), or OpenOffice Calc (free) are huge time-savers.

Only a very basic understanding of spreadsheets is required to successfully build your own calculation tool. As such, we will assume that nearly all readers will choose to use a spreadsheet program to perform their feasibility calculations. To make it super easy for you to follow along, each step is presented as a spreadsheet template. Basically, we'll show you exactly how to build your own spreadsheet table and which equations to put in which cells. We will also present all equations in the spreadsheet "language" of formulas, functions, cell references, and operators.

Regardless of your comfort with spreadsheet software, if you follow the templates you'll end up with a powerful calculation tool that will save you an incredible amount of time when testing iterations and design scenario permutations.

Don't want to build your own tool? No problem. Head to www.essentialrwh.com if you'd like to hit the ground running by purchasing our spreadsheet-based *Essential Rainwater Harvesting Tool*.

We can't really imagine doing these calculations without a spreadsheet, but it is doable. If you prefer to do the math by hand, you'll absolutely be able to follow along with the instructions and formulas provided. In that case, a simple calculator is all that you'll need.

A Spreadsheet Cheatsheet

Need a refresher on how to use a spreadsheet? Almost everything you need to know to build your own RWH calculation tool is summarized below:

Formulas: Formulas always start with an equal sign (=), and they contain any or all of the following: functions, cell references, and operators.

Functions: Functions are predefined mathematical operations. We'll use:

- SUM: adds the values in cells.
- IF: allows you to make logical comparisons.

Cell References: This is the identification of a particular cell. In the templates presented in this book, we use the same convention as in all spreadsheet programs. Columns are labeled with a letter (A, B, C, D...) and rows with numbers (1,2,3,4...). Cells are referenced by their column letter and row number. For instance, cell C2 is the cell represented by column C, row 2.

We may also reference a range of cells using the : (colon) symbol. Cell C2:C5 means cells C2, C3, C4, and C5. Alternatively A28:C28 means cells A28, B28, and C28.

Also, when you are building equations in a spreadsheet you'll save yourself a ton of time if you are familiar with how to use relative cell references and absolute references (using the $ sign).

Operators: The operators used in spreadsheets (and, consequently, those used in the templates here) are: +,-,*,/, >,<, and ^. Note that the * (asterisk) operator is a multiplier, the / (slash) operator is used for division. The < (less-than sign) and the > (greater-than sign) are used for comparisons, often within IF functions. The ^ symbol is used to denote an exponential operator. For instance 10^3 is 10*10*10.

Tabs: Tabs are simply separate "worksheets" within the application; they help to keep information and calculations organized. You can think of each of your tabs as pieces of paper within a stack.

If you have very little experience using spreadsheets, or if you are unfamiliar with how cell references work (using the $ sign), simply search for a basic online tutorial for the program of your choice (typically Microsoft Excel, Mac Numbers, Google Sheets, or OpenOffice Calc), and you'll be able to get started in no time.

	A	B	C	D	E	F
1						
2						
3						
4						
5						

Chapter 3

Feasibility

THERE'S A FIRST MAJOR STAGE to solving any design problem: the feasibility stage. RWH system design is no different.

In the feasibility stage of a RWH design, the system is defined as the two major components: your roof and your tank. These are combined with your performance goals and demand needs. You are simply trying to understand if there is a combination of roof area and tank volume that is technically feasible and will work for you and your budget.

Establishing your system size and defining volumes is also particularly useful if you would like to perform any economic calculations, such as determining the water-supply fees that you might offset with your RWH system, or if you are comparing several options for your water supply. The feasibility calculations will aid you in attempting to make any such cost evaluations or projections.

The Three Parts of the Feasibility Stage

We present the feasibility stage of RWH system design in three parts:

Part 1: Define Supply and Demand

Part 2: Optimize Your Storage

Part 3: Test and Redesign for Low Rainfall

In Part 1, you start by collecting the rainfall data for your location to determine the rainwater supply in average conditions. You'll also define and calculate the volume of your rainwater demands.

In Part 2, you'll look at the average rainwater supply and your demands over the calendar year and you'll determine the optimum storage capacity that ensures that you don't run out of water at any point in the year.

In Part 3, you'll take the system designed in Part 2 for average rainfall conditions, and you'll evaluate how it would perform with less than average rain. Depending on your risk tolerance and minimum operating goals, you'll likely make adjustments to your system design at this stage.

All three parts are separate but interconnected iterative exercises. The process often requires you to seek an input for a given output. We'll go into more detail later, but this is why attempting to perform these calculations by hand is so impractical.

The critical order of design parameters is: (1) reduce/optimize demand; (2) increase/optimize roof area; and (3) optimize tank volume. When iterating for a solution, always start with critically evaluating and/or (re)designing your water usage. Next, evaluate your ability to increase roof area, and lastly, only increase your storage volume if you can't do the first two.

In the following sections, we'll provide templates that will show you exactly how to build your own feasibility tool using spreadsheet software. There are templates for each of the three parts. Once you've created your spreadsheet tool, we'll teach you how to use it, step-by-step. Also, in the same sense that *a picture is worth a thousand words,* we opted for: *a template is worth a thousand words.* Rather than describe with words the mathematical operations and logic for each step and calculation, the templates show the exact formulas to put in each cell.

Again, if you don't want to use the templates provided to build the spreadsheet tool for yourself, head to www.essentialrwh.com to purchase our spreadsheet-based Essential Rainwater

Harvesting Tool. If you do pick up our tool, you'll still want to the follow the processes for using the tool presented here.

All Models Are Wrong...

> The most that can be expected from any model is that it can supply a useful approximation to reality: All models are wrong; some models are useful.
>
> — George Box, Statistician

Before diving in to the calculation tool and processes, some caveats are required.

Performing sizing calculations (and any subsequent economic evaluations) on RWH systems is based on building a mathematical model that is, at best, an approximation or a simplification of reality. Lucas et al. (2006) compared spreadsheet modeling to more comprehensive simulation software and showed that spreadsheet modeling underestimated supply and overestimated the required size of storage. The difference was particularly significant (up to 20%) in tank sizes smaller than 2,000 liters (530 gal). Above 2,000 liters (530 gal), the differences between models converged within ± 10%. Also important to note is that the spreadsheet model used in their analysis was based on the use of a daily time step. (More on this, later.)

The spreadsheet calculation tool presented in this book uses an even broader simplification — a *monthly* time step. We opted for this simplification for two major reasons:

1. For the average homeowner, daily rainfall data is not nearly as easy to get as is monthly rainfall data.

2. The increased amount of data needed for a daily time step is cumbersome for the average user and would have added substantial complexity to the spreadsheet templates described and provided in this book.

In order to compensate for this simplification (and, particularly, to compensate for the risks it might pose for the RWH system user), our feasibility evaluation includes a very important component: Part 3. The intention with Part 3 is that you take your design (based, as it is, on simplified assumptions), and you look for *fragility*. Based on your risk tolerance, you can then add in varying levels of overall robustness and resilience to your system design.

We feel comfortable that the monthly time-step spreadsheet model combined with the process presented in Part 3 is reasonable for a majority of home-scale RWH system users/designers — especially those who are not basing their decision to install a RWH system purely on traditional financial metrics. These users know that their RWH system has a greater return on investment when personal values are taken into account (sustainability, resilience, etc.).

On the other hand, this simplified approach may not be appropriate or robust enough for certain evaluations, such as a municipality looking to determine the actual cost savings in large-scale adoption of home-scale RWH systems. A municipality wanting to encourage homeowners to install RWH systems needs to accurately understand the economics, affordability, and benefits of such a program. This would necessitate more sophisticated modeling and simulation tools.

If you prefer to be less conservative in your calculations, note that the spreadsheet templates provided here can easily be modified to use daily time step, instead of monthly time step. You would need to find and input 365 rows of rainfall data (Jan 1, Jan 2, Jan 3, and so on) instead of 12 (Jan, Feb, March, and so on). You'd also have 365 rows of calculations instead of 12. The calculation methodology and the step-by-step process for Part 1, Part 2, and Part 3 would remain unchanged, regardless of the time step used. And really, if you were up for the data management challenge, and could get the rainfall and demand data on a very high level of granularity, there's no reason you couldn't perform your calculations on an even smaller time step, say, every 5 minutes. Our Essential Rainwater Harvesting Tool includes the ability to perform optimization calculations in two time-step options: daily or monthly.

We do know of initiatives to create relatively simple algorithms that may one day be used to provide more accurate analysis for the average home-scale RWH user and designer. We will keep our website up to date with such initiatives as they become available.

Part 1. Define Supply and Demand

The Part 1 Templates and Processes

The step-by-step process of the first part of the feasibility stage is shown in Figure 3.1. The Part 1 process contains Steps 1.1 through to 1.7. After you've gone through all seven steps, you cannot continue onto Part 2 until you pass the Supply/Demand test (the bolded diamond). The spreadsheet templates for each step as well as the detailed description of the Part 1 process are provided in the sections that follow.

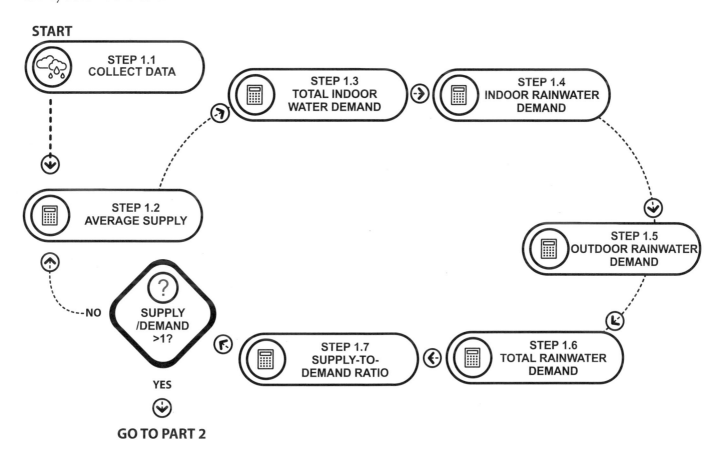

Fig. 3.1: *The step-by-step process for Part 1 of the feasibility stage: Define Supply and Demand.*

CREDIT: VERGE PERMACULTURE/S. FIDLER

**STEP 1.1
COLLECT DATA**

Step 1.1: Collect Data

Step 1.1 Template

The first thing you'll do is gather all of the relevant data you'll need to get started with your design. Pull all of your information together on the first tab of your spreadsheet by building a table just like the one shown in the Step 1.1 Template.

Relevant Location Information: Describe the project location (such as an address). Use the nearest weather station and the nearest major city. Although they are often the same, many major cities have several weather stations. If that is the case, select the station at the airport, as it's often the most complete dataset.

Climate: The appropriate government agency for your country is usually the first place you'll turn to for weather and rainfall data. In the U.S., the National Centers for Environmental Information (www.ncdc.noaa.gov) and in

Canada, Environment Canada (climate.weather. gc.ca) publish *Climate Normals* for all major cities every ten years. Climate Normals are 30-year datasets that describe the average climatic conditions for a particular location. This includes averages for temperature, rainfall, snowfall, precipitation, wind speed, and, depending on the weather station, they may also include humidity, pressure, and radiation.

Most jurisdictions make the data available online, so start by searching your country name followed by "Climate Normals," and you'll likely find the data, if it is available. For instance, an online search using the term "Australia Climate Normals" brings up the Australian Bureau of Meteorology. From there, it is fairly straightforward to navigate to the Climate Normals page.

Once you pull up the Climate Normals for the nearest weather station/city, you'll want to record the following information.

- Data range for your Climate Normals (yyyy–yyyy).
- Average rainfall in mm (in) for each month.

Step 1.1 Template: Collect data and Climate Normals on a month-by-month basis.

	A	B	C	D	E	F	G	H	I	J	K	L	M	N	O
1		Location Description:													
2		Weather Station/City:													
3		Climate Data Range:													
4															
5			Units	Jan	Feb	Mar	Apr	May	Jun	Jul	Aug	Sep	Oct	Nov	Dec
6	Rain	Average rainfall													
7	Temp	Average by month													
8		Daily maximum													
9		Extreme min													
10		Extreme max													
11	Snow	Snowfall													
12		Average snow depth													
13	Wind	Most frequent direction													
14		Average speed													
15		Maximum gust speed													
16	Humidity	Average relative humidity													

- Total annual rainfall in mm (in). This may be provided, or you may need to calculate it by simply adding up the monthly average rainfall numbers.

Be aware that your jurisdiction might mandate the rainfall data you have to use in your system design — another reason you must always start by understanding your regulatory context.

Although not necessarily directly needed for system calculation, we recommend that while you are there, you also grab some of the following very useful data that will help you understand your particular climatic patterns and inform other parts of your system design:

- **Temperature in °C (°F):** The monthly temperature data provides design insight, particularly if you are irrigating a landscape or designing a system in a cold climate and need to consider freeze protection. You'll want to collect several datasets: average by month, daily maximum, extreme minimum, and extreme maximum.
- **Snow in cm (in):** Snowfall data is useful if you are considering harvesting snow to increase the amount of captured water. Note that we've seen very few cases where it actually makes sense to harvest snow (and it adds substantial complexity to a system). As such, it is not covered in depth in this book.
- **Wind in m/s (ft/s):** Average wind speed, maximum wind gust speed, and most frequent direction may provide insight into how you need to fasten your system components, particularly if your tank is a lightweight material, such as plastic.
- **Relative Humidity in %:** If you are looking at providing outdoor irrigation with your rainwater, you may want to grab relative humidity, or at least the average midsummer high relative humidity. Later on, we'll provide a way to quickly estimate a rough volume for irrigation using this number.
- **Data for Cold Climates:** For seasonal systems in cold climates, we recommend that you also collect the first and last frost date of the season. These dates will help you to decide approximately when you should plan on commissioning and decommissioning your system every year.

EXAMPLE

Example: Collect data for an urban home in Vancouver, British Columbia, Canada (Step 1.1)

A family in Vancouver, British Columbia, is investigating the feasibility of a RWH system for their new home. The first step of the feasibility stage is *Step 1.1: Collect Data,* so they open a new spreadsheet, create the first tab called "Step 1.1," and build a blank table following the Step 1.1 Template.

They then navigate to the Environment Canada webpage, at climate.weather.gc.ca, and select *Canadian Climate Normals.* A search for the station name "Vancouver" displays six different weather stations. They select the Vancouver International Airport

and navigate to *Normals Data.* They copy the desired information from this webpage to their spreadsheet and/or they export the data in a .csv file, which can then be opened in spreadsheet software.

They input all of their data into their spreadsheet and the completed table looks like this:

For a video tutorial version of this example head to www.essentialrwh.com

Step 1.1 Example: Metric Units

	A	B	C	D	E	F	G	H	I	J	K	L	M	N	O
1		Location Description:			Vancouver, BC										
2		Weather Station/City:			Vancouver International Airport										
3		Climate Data Range:			1971–2000										
4															
5			Units	Jan	Feb	Mar	Apr	May	Jun	Jul	Aug	Sep	Oct	Nov	Dec
6	Rain	Average rainfall	mm	139	114	112	84	68	55	40	39	54	113	179	161
7	Temp	Average by month	deg C	3.3	4.8	6.6	9.2	12.5	15.2	17.5	17.6	14.6	10.1	6.0	3.5
8		Daily maximum	deg C	6.1	8.0	10.1	13.1	16.5	19.2	21.7	21.9	18.7	13.5	9.0	6.2
9		Extreme min	deg C	-17.8	-16.1	-9.4	-3.3	0.6	3.9	6.7	6.1	0.0	-5.9	-14.3	-17.8
10		Extreme max	deg C	15.3	18.4	19.4	25.0	30.4	30.6	31.9	33.3	29.3	23.7	18.4	14.9
11	Snow	Snowfall	cm	16.6	9.6	2.6	0.4	0.0	0.0	0.0	0.0	0.0	0.1	2.5	16.3
12		Average snow depth	cm	1.0	1.0	0.0	0.0	0.0	0.0	0.0	0.0	0.0	0.0	0.0	1.0
13	Wind	Most frequent direction	-	E	E	E	E	E	E	E	E	E	E	E	E
14		Average speed	km/h	11.5	12.1	12.9	12.6	12.0	11.7	11.5	11.0	10.6	11.0	12.3	12.0
15		Maximum gust speed	km/h	97	119	108	100	90	70	71	85	91	126	129	100
16	Humidity	Average relative humidity	%	79.8	75.3	70.2	65.4	63.9	63.6	62.4	63.1	67.8	75.8	78.9	80.8

Step 1.1 Example: US Customary Units

	A	B	C	D	E	F	G	H	I	J	K	L	M	N	O
1		Location Description:		Vancouver, BC											
2		Weather Station/City:		Vancouver International Airport											
3		Climate Data Range:		1971–2000											
4															
5			Units	Jan	Feb	Mar	Apr	May	Jun	Jul	Aug	Sep	Oct	Nov	Dec
6	Rain	Average rainfall	in	5.5	4.5	4.4	3.3	2.7	2.2	1.6	1.5	2.1	4.4	7.0	6.3
7	Temp	Average by month	deg F	37.9	40.6	43.9	48.6	54.5	59.4	63.5	63.7	58.3	50.2	42.8	38.3
8		Daily maximum	deg F	43.0	46.4	50.2	55.6	61.7	66.6	71.1	71.4	65.7	56.3	48.2	43.2
9		Extreme min	deg F	0.0	3.0	15.1	26.1	33.1	39.0	44.1	43.0	32.0	21.4	6.3	0.0
10		Extreme max	deg F	59.5	65.1	66.9	77.0	86.7	87.1	89.4	91.9	84.7	74.7	65.1	58.8
11	Snow	Snowfall	in	6.5	3.8	1.0	0.2	0.0	0.0	0.0	0.0	0.0	0.0	1.0	6.4
12		Average snow depth	in	0.4	0.4	0.0	0.0	0.0	0.0	0.0	0.0	0.0	0.0	0.0	0.4
13	Wind	Most frequent direction	-	E	E	E	E	E	E	E	E	E	E	E	E
14		Average speed	mph	7.1	7.5	8.0	7.8	7.5	7.3	7.1	6.8	6.6	6.8	7.6	7.5
15		Maximum gust speed	mph	60	74	67	62	56	44	44	53	57	78	80	62
16	Humidity	Average relative humidity	%	79.8	75.3	70.2	65.4	63.9	63.6	62.4	63.1	67.8	75.8	78.9	80.8

**STEP 1.2
AVERAGE SUPPLY**

Step 1.2: Average Supply
Collection Area

It's time to differentiate between *roof* and *catchment area*. The roof is the covering on your home; in North American architecture, this is typically sloped. The catchment area is, in a sense, how the sky sees your roof from above. It is the area available for capturing rain, and it is the roof area projected vertically, as shown in Figure 3.2. You can see that regardless of the complexity, the slope, or shape of the roof, the catchment area is quite simple to calculate, and it will always

be smaller than the actual roof area (unless you have a flat roof). In Figure 3.2, if the house length was 20 m (65.6 ft) and the house width was 10 m (32.8 ft), then the catchment area would be 200 m² (2,153 ft²).

Roof Efficiency

Much of the other literature on designing RWH systems includes a factor called *roof efficiency;* this is sometimes called *the runoff coefficient.* The idea is that this coefficient describes the fraction of water that hits your roof but doesn't make it to your storage. Efficiency values are often stated based on the choice of roof material, which implies losses due to material porosity or absorption.

However, the reality is that unless you choose a porous roof material (like a green roof) or unless you have ponding of water on the roof

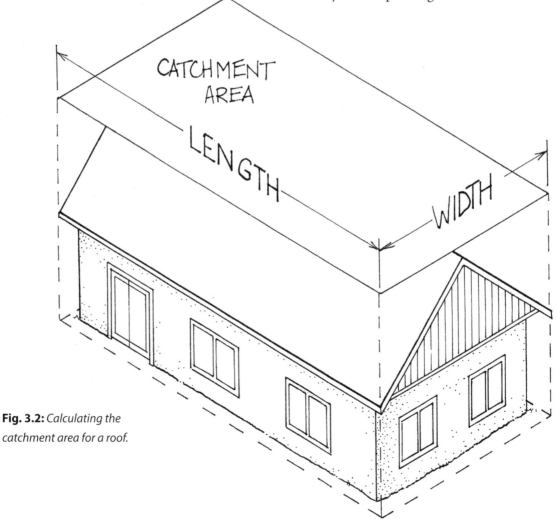

Fig. 3.2: *Calculating the catchment area for a roof.*

surface itself (due to flat spots or inappropriately designed roof slope), any significant water losses are far more likely to be related to leaking gutters and gutter overflows. Improperly designed and installed gutters have been shown to cause losses as high as 20%.

In Table 3.1, we re-publish typical roof efficiencies and suggest that you use these as a guideline. If you have leaky or poorly installed gutters, or a very low-pitched roof with ponding issues, consider reducing the efficiency figure even more than provided. Alternatively, if you carefully select nonporous and appropriate roofing material (such as metal) and properly design and install your roof detail and gutters to eliminate losses, you could arguably use a roof efficiency near 100%.

Step 1.2 Template

The equation used to calculate the amount of water that can be harvested from your roof is:

Metric: Supply (liters) = catchment area (m²) × roof efficiency × rainfall (mm)

Table 3.1

Roofing Material	Efficiency
Metal	0.95
Asphalt	0.90
Clay / Concrete Tiles / Slate	0.90
EPDM Rubber Roofing, PVC etc.	0.95–0.99
Tar and Gravel (flat)	0.80–0.85

Table 3.1:
Roof efficiency based on material type.
ADAPTED FROM: Novak, *DESIGNING RAINWATER HARVESTING SYSTEMS,* JOHN WILEY & SONS, 2014.

US Customary: Supply (gal) = catchment area (ft²) × roof efficiency × rainfall (in) × 0.623

For Step 1.2 you'll want to build a table to calculate the supply on a month-by-month basis (in liters/yr [gal/yr]) as well as sum up the individual month volumes to arrive at your Total Average Supply. The equations are shown for both metric and US customary units in the Step 1.2 template.

If you are in a cold climate and only doing seasonal rainwater harvesting, you'll only include those months in which you intend on harvesting rainwater, typically the spring, summer, and fall.

Step 1.2 Template: Average Supply

	A	B	C	
1	Roof Catchment, m² [ft²]			
2	Roof Efficiency[τ]			
3				
4	Month	Average Rainfall[ττ]	Supply	
5		mm [in]	*Metric Formula, litres*	*US Cust. Formula, gal*
6	Jan	=Step 1.1 cell D6	=B1*B2*B6	=B1*B2*B6*0.623
7	Feb	=Step 1.1 cell E6	=B1*B2*B7	=B1*B2*B7*0.623
8	Mar	=Step 1.1 cell F6	=B1*B2*B8	=B1*B2*B8*0.623
9	Apr	=Step 1.1 cell G6	=B1*B2*B9	=B1*B2*B9*0.623
10	May	=Step 1.1 cell H6	=B1*B2*B10	=B1*B2*B10*0.623
11	Jun	=Step 1.1 cell I6	=B1*B2*B11	=B1*B2*B11*0.623
12	Jul	=Step 1.1 cell J6	=B1*B2*B12	=B1*B2*B12*0.623
13	Aug	=Step 1.1 cell K6	=B1*B2*B13	=B1*B2*B13*0.623
14	Sep	=Step 1.1 cell L6	=B1*B2*B14	=B1*B2*B14*0.623
15	Oct	=Step 1.1 cell M6	=B1*B2*B15	=B1*B2*B15*0.623
16	Nov	=Step 1.1 cell N6	=B1*B2*B16	=B1*B2*B16*0.623
17	Dec	=Step 1.1 cell O6	=B1*B2*B17	=B1*B2*B17*0.623
18	Total Average, liters/yr [gal/yr]	=sum(B6:B17)	=sum(C6:C17)	

τ - *From Table 3.1*
ττ - *From average rainfall data (Step 1.1)*

EXAMPLE

Example: Rainwater supply calculation for an urban home (Step 1.2)

Continuing on from the previous example: The house measures 20 m (65.6 ft) by 10 m (32.8 ft), and the family is already anticipating installing a metal roof.

On Table 3.1 they find a roof efficiency of 0.95.

To calculate catchment area:

Catchment area in m² (ft²) = 20 m (65.6 ft) * 10 m (32.8 ft) = 200 m² (2,153 ft²)

They create a second tab in their spreadsheet, called "Step 1.2," and replicate the Step 1.2 Template. They use formulas to grab the rainfall data in the Step 1.1 table and input this data into Step 1.2 column B, as shown below. Using the supply formula in the units of their choice (both are shown), they calculate the month-by-month supply in column C.

Their Total Average Supply based on their current catchment area is 219,393 liters/yr (57,979 gal/yr).

Step 1.2 Example

	A	B		C		
1	Roof Catchment, m² [ft²]	200 [2153]				
2	Roof Efficiency$^\tau$	0.95				
3						
4	Month	Average Rainfall$^{\tau\tau}$		Supply		
5		mm [in]	*Formula*^{Y}	liters [gal]	*Metric Formula^Y, litres*	*US Cust. Formula^Y, gal*
6	Jan	139.1 [5.5]	='Step 1.1'!D6	26,429 [7,008]	*=B1*B2*B6*	*=B1*B2*B6*0.623*
7	Feb	113.8 [4.5]	='Step 1.1'!E6	21,622 [5,734]	*=B1*B2*B7*	*=B1*B2*B7*0.623*
8	Mar	111.8 [4.4]	='Step 1.1'!F6	21,242 [5,607]	*=B1*B2*B8*	*=B1*B2*B8*0.623*
9	Apr	83.5 [3.3]	='Step 1.1'!G6	15,865 [4,205]	*=B1*B2*B9*	*=B1*B2*B9*0.623*
10	May	67.9 [2.7]	='Step 1.1'!H6	12,901 [3,440]	*=B1*B2*B10*	*=B1*B2*B10*0.623*
11	Jun	54.8 [2.2]	='Step 1.1'!I6	10,412 [2,803]	*=B1*B2*B11*	*=B1*B2*B11*0.623*
12	Jul	39.6 [1.6]	='Step 1.1'!J6	7,524 [2,039]	*=B1*B2*B12*	*=B1*B2*B12*0.623*
13	Aug	39.1 [1.5]	='Step 1.1'!K6	7,429 [1,911]	*=B1*B2*B13*	*=B1*B2*B13*0.623*
14	Sep	53.5 [2.1]	='Step 1.1'!L6	10,165 [2,676]	*=B1*B2*B14*	*=B1*B2*B14*0.623*
15	Oct	112.5 [4.4]	='Step 1.1'!M6	21,375 [5,607]	*=B1*B2*B15*	*=B1*B2*B15*0.623*
16	Nov	178.5 [7]	='Step 1.1'!N6	33,915 [8,920]	*=B1*B2*B16*	*=B1*B2*B16*0.623*
17	Dec	160.6 [6.3]	='Step 1.1'!O6	30,514 [8,028]	*=B1*B2*B17*	*=B1*B2*B17*0.623*
18	Total Average, liters/yr [gal/yr]	1,155 [45.5]	=sum(B6:B17)	219,393 [57,979]	*=sum(C6:C17)*	

τ - From Table 3.1
$\tau\tau$ - From average rainfall data (Step 1.1)
Y - The formulas contained within the cells are shown, but they would not be visible as such in your spreadsheet
Calculations in each choice of unit were performed independently and values in earlier input fields rounded to a significant number of digits. The end result is a rounding discrepancy if you compare the metric vs US Customary results.

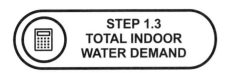

**STEP 1.3
TOTAL INDOOR
WATER DEMAND**

Step 1.3: Total Indoor Water Demand

Unless you have absolutely no intention of supplying indoor water with rainwater, the next step is to calculate the total water demands for your home, including even the non-rainwater uses.

Your total indoor water volume will help you make better economic and risk decisions about backup and cost of supply and could certainly be a major factor influencing the sizing and supply risk considerations for your RWH system. Also, if you are on an acreage, you'll certainly need this information to size and design any septic or onsite wastewater system.

Typical indoor water usage volumes for North American homes are shown in Table 3.2.

Table 3.2: Typical Indoor Fixtures and Water Usage

Fixtures	Fixture Type	Average Manufacturer's Rating			Average Usage Pattern			Average Daily Indoor Water Volume
		Lpf [gpf]	Lpm [gpm]	liters/load [gallons/load]	total flushes/person/day[Y]	mins/person/day	loads/person/day	Lpd/person [gpd/person]
Toilet	Inefficient/old	15.1 [4]			5.1			77 [20.4]
Toilet	Low flush	13.2 [3.5]			5.1			67.3 [17.9]
Toilet	Ultra-low flush	6.1 [1.6]			5.1			31.1 [8.2]
Toilet	Dual flush	4.9 [1.3]			5.1			25 [6.6]
Shower	Inefficient/old		9.5 [2.5]			5.3		50.4 [13.3]
Shower	Standard		8.3 [2.2]			5.3		44 [11.7]
Shower	High-efficiency		5.7 [1.5]			5.3		30.2 [8]
Faucets: Kitchen + Lavatory	Inefficient/old		6.3 [1.66]			8.1		51 [13.4]
Faucets: Kitchen + Lavatory	Standard		5.1 [1.34]			8.1		41.3 [10.9]
Faucets: Kitchen + Lavatory	High-efficiency		3.8 [1]			8.1		30.8 [8.1]
Faucets: Lavatory only	Inefficient/old		6.3 [1.66]			1.5		9.5 [2.5]
Faucets: Lavatory only	Standard		5.1 [1.34]			1.5		7.7 [2]
Faucets: Lavatory only	High-efficiency		3.8 [1]			1.5		5.7 [1.5]
Dishwasher	Standard			17 [4.5]			0.24	4.1 [1.1]
Dishwasher	High-efficiency			13.2 [3.5]			0.24	3.2 [0.8]
Laundry	Top Loading			151 [40]			0.37	55.9 [14.8]
Laundry	Front Loading, Standard			102 [27]			0.37	37.7 [10]
Laundry	Front Loading, High-efficiency			τ			0.37	$\tau\tau$

Lpf = Liters per flush
gpf = gallons per flush
Lpm = Liters per minute
gpm = gallons per minute
mins/person/day = average total number of minutes per day the fixture is used per person

Lpd = Liters per day
gpd = gallons per day
τ - See manufacturer's rating
$\tau\tau$ - Calcuate by multiplying the manufacturer's rating by the usage pattern
Y - All household toilets combined

Table 3.2: *Typical indoor water usage in North American homes based on average manufacturer's rating and average usage patterns.*

Source: Adapted from Vickers, A. 2001. Handbook of Water Use and Conservation. Water Flow Press. Amherst, MA.

Step 1.3 Template

In Step 1.3 (your third tab), you'll create a table for your indoor fixtures and anticipated usage patterns. Here is the Step 1.3 template:

Step 1.3 Template: Total Indoor Water Demand

	A	B	C	D	E	F	G	H	I
1			Manufacturer's Rating$^\tau$			Usage Pattern$^{\tau\tau}$			Daily Indoor Water Volume
2	Fixtures	Fixture Type	Lpf [gpf]	Lpm [gpm]	liters/load [gallons/load]	total flushes/person/day	mins/person/day	loads/person/day	Lpd/person [gpd/person]
3	Fixture 1 (e.g. Toilet 1)	Dual flush							=C3*F3
4	Fixture 2 (e.g. Toilet 2)	Dual flush							=C4*F4
5	Fixture 3 (Shower)	High-efficiency							=D5*G5
6	Fixture 4 (e.g. Faucets: Kitchen + Lavatory)	High-efficiency							=D6*G6
7	Fixture 5 (e.g. Dishwasher)	High-efficiency							=E7*H7
8	Fixture 6 (e.g. Laundry)	Front Loading, High-efficiency							=E8*H8
9									
10							Total of All Fixtures, Lpd/person [gpd/person]		=sum(I3:I9)
11							Number of Occupants		
12							Total Daily Indoor Water Demand, Lpd [gpd]		=I10*I11
13							Monthly Indoor Water Demand, L/month [gal/month]		=I12*30.4^Y

τ -Taken from actual manufacturer's rating or from Table 3.2.
$\tau\tau$ - Taken from actual water audit or from Table 3.2.
Y - 30.4 is the average number of days per month in a year (365/12)

In rows 3–9 (include as many as you may need), you'll list all of the water-supply fixtures in your home. Input the manufacturer's rating and usage patterns as appropriate in columns C–H.

Here are some tips when filling in and building the Step 1.3 template for your particular scenario:

- Column C, D, and E are volumes based on the physical design of the fixture and come from the manufacturer. If you don't have manufacturer's data, you can use the typical average values in Table 3.2 as estimates.

- Columns F, G, and H are entirely based on habits and usage patterns of the occupants. The numbers provided in Table 3.2 are based on average toilet, showering, dishwasher, and washing machine use in a typical North American household.

- For each fixture type, you take the manufacturer's rating (column C, D, or E) and multiply it by the appropriate usage pattern (column F, G, or H) to estimate your Daily Indoor Water Volume (column I).

- If you know the actual manufacturer's rating for the fixtures you are using (column C, D, or

E), you can use those instead for a more accurate estimate. Know that if you are using the manufacturer's rating for your faucets, you can de-rate the maximum flow rate by about two-thirds because faucets are not usually opened full throttle unless you are filling a container.

- Washing machines and dishwasher ratings (column E) can vary significantly. If possible, use the actual manufacturer's rating instead of the typical manufacturer's ratings provided. If you don't know the rating, choose something that is conservative.

- If you use 5.1 flushes per person per day (from Table 3.2) for column F, remember that this estimate is based on all household toilets combined. Therefore, to estimate the number of flushes an individual toilet may receive (per person) divide this number by the total number of toilets in the house.

- If you do an actual water audit of your home and know what your actual habits and patterns of usage are, there's an opportunity to reduce the values in columns F, G, and H.

In column I of the Step 1.3 Template, the formulas to calculate the Daily Indoor Water Volume per fixture, the Total of All Fixtures, the Total Daily Indoor Water Demand, and the Monthly Indoor Water Demand are shown.

Step 1.4: Indoor Rainwater Demand
Step 1.4 Template
Your Indoor Rainwater Demand is simply the amount of the indoor water that you intend on supplying with rainwater.

We've seen a lot of scenarios where the use of rainwater is regulated to such an extent that the possibilities for indoor uses are very limited. For instance, in Alberta, Canada, only toilet/urinal flushing is legally allowed. To design a system that meets the legal requirements in Alberta (without getting approval through an alternative compliance pathway), your Indoor Rainwater Demand would be limited from the onset to the toilet fixtures in your home.

In Step 1.4, you'll repeat the exercise given in Step 1.3, but only with those fixtures that you intend on using with rainwater.

Note that we want you to calculate your Indoor Rainwater Demand as a percentage of your Total Indoor Water Demand (cell I33). To do this, you'll need to go back and grab the Daily Indoor Water Demand from Step 1.3, cell I12. Again, the beauty of a using a spreadsheet is that you can simply reference the cell in Step 1.3, as shown in the formula provided in the template. If any subsequent changes are made to your tables in Steps 1.1–1.3, the calculation in Step 1.4 is automatically updated.

If you are in a scenario where you intend to provide 100% of your indoor water needs with rainwater, you can skip the Step 1.4 fixture-by-fixture calculation (rows 23–28, in our example) and override the formula in I33 by inputting 100%. Your Monthly Indoor Rainwater Demand (I34) will automatically calculate as being equal to your Monthly Indoor Water Demand, from Step 1.3.

Alternatively, if from the onset your goal is to only supplement your indoor water (vs provide rainwater to specific fixtures), you could be thinking: *I'd love to supply 20% of my total indoor*

Step 1.4 Template: Indoor Rainwater Demand

	A	B	C	D	E	F	G	H	I
21			Manufacturer's Rating[τ]			Usage Pattern[ττ]			Daily Indoor Water Volume
22	Rainwater Fixtures	Fixture Type	Lpf [gpf]	Lpm [gpm]	liters/load [gallons/load]	total flushes/person/day	mins/person/day	loads/person/day	Lpd/person [gpd/person]
23	Toilet 1	Dual flush							=C23*F23
24	Toilet 2	Dual flush							=C24*F24
25									
26									
27									
28	Laundry	Front Loading, High-efficiency							=E28*H28
29									
30	Total of All Rainwater Fixtures, Lpd/person [gpd/person]								= sum(I23:I29)
31	Number of Occupants								=Step 1.3 cell I11
32	Total Daily Indoor Rainwater Demand, Lpd [gpd]								=I30*31
33	Percentage of Indoor Water to Be Supplied with Rainwater,%								=I32/(Step 1.3 cell I12)
34	Monthly Indoor Rainwater Demand, L/month [gal/month]								=I33* (Step 1.3 cell I13)

τ -Taken from actual manufacturer's rating or from Table 3.2.

ττ - Taken from actual water audit or from Table 3.2.

water demands with rainwater. Again, override the formula in I33 with 20%, and calculate your Monthly Indoor Rainwater Demand (I34) based on 20% of your Monthly Indoor Water Demand, from Step 1.3.

The reality is that you might not know at this point what actual percentage of rainwater use in your home is realistic, economic, or feasible. That's okay — this is why you are at the feasibility stage. Simply get started with an estimate. Soon, you'll be testing the viability of your Total Indoor Rainwater Demand, and iterating it, if necessary.

 **EXAMPLE**

Example: Total indoor water demand and indoor rainwater demand (Step 1.3 and Step 1.4.)

The family from our example is planning on installing the following fixtures and fixture-types in their home: two dual-flush toilets, high-efficiency shower, high-efficiency faucets throughout, a high-efficiency dishwasher, and a front-loading high-efficiency laundry machine with a manufacturer's rating of 76 liters/load (20 gal/load).

They have already made the decision that they intend on only attempting to supply toilets and their laundry with rainwater. The remainder of their indoor water will be supplied by the municipal system.

To calculate the water use in their home, they start a new tab in their spreadsheet, and then build two new tables based on

Step 1.3 Example: Total indoor water demand

	A	B	C	D	E	F	G	H	I	
1			Manufacturer's Rating[τ]			Usage Pattern[ττ]			Daily Indoor Water Volume	
2	Fixtures	Fixture Type	Lpf [gpf]	Lpm [gpm]	liters/ load [gallons/ load]	total flushes/ person/ day	mins/ person/ day	loads/ person /day	Lpd/ person [gpd/ person]	Formula[ξξ]
3	Toilet 1	Dual flush	4.9 [1.3]			2.55[YY]			12.5 [3.3]	=C3*F3
4	Toilet 2	Dual flush	4.9 [1.3]			2.55[YY]			12.5 [3.3]	=C4*F4
5	Shower	High-efficiency		5.7 [1.5]			5.3		30.2 [8]	=D5*G5
6	Faucets: Kitchen + Lavatory	High-efficiency		5.7 [1]			5.3		30.2 [8.1]	=D6*G6
7	Dishwasher	High-efficiency			13.2 [3.5]			0.24	3.2 [0.8]	=E7*H7
8	Laundry	Front Loading, High-efficiency			76 [20.1]			0.37	28.1 [7.4]	=E8*H8
9										
10						Total of All Fixtures, Lpd/person [gpd/person]			117 [28]	= sum(I3:I9)
11						Number of Occupants			4	
12						Total Daily Indoor Water Demand, Lpd [gpd]			468 [112]	=I10*I11
13						Monthly Indoor Water Demand, L/month [gal/month][ξ]			14,227 [3,405]	= I12*30.4[Y]

τ -Taken from actual manufacturer's rating or from Table 3.2.
ττ - Taken from actual water audit or from Table 3.2.
Y - 30.4 is the average number of days per month in a year (365/12)
ξ - Domestic water usage assumed to be constant throughout the year (i.e. over 12 months)

YY - Lpd/person from Table 3.2 divided by 2 to account for 2 toilets in the home
ξξ - *The formulas contained within the cells are shown, but they would not be visible as such in your spreadsheet*

the Step 1.3 and Step 1.4 Templates. They estimate the manufacturer's ratings and their own water usage based on Table 3.2, with the exception of their laundry machine, which they pull directly from the manufacturer's specifications.

The formulas, as well as the results, are shown in Column I. Their Daily Indoor Rainwater Demand (Step 1.4 cell I32) is 212 liters/day (56 gal/day) and their Total Indoor Water Demand (Step 1.3 cell I12) is 468 liters/day (112 gal/day). That means that they are going to try and supply 45% of their Total Indoor Water Demand with rainwater.

(*Note to reader:* This is a perfect example of the rounding discrepancies mentioned earlier in the book that are due to presenting in multiple units. The authors realize that 468 liters/day is actually 128 gal/day, not 112 gal/day. However, when US Customary units are used in earlier input fields, and these are rounded to the nearest significant number of digits, the end result is a small rounding discrepancy.)

Step 1.4 Example: Indoor rainwater demand from toilets and laundry

	A	B	C	D	E	F	G	H	I	
21			Manufacturer's Rating[τ]			Usage Pattern[ττ]			Daily Indoor Water Volume	
22	Fixtures	Fixture Type	Lpf [gpf]	Lpm [gpm]	liters/ load [gallons/ load]	total flushes/ person/ day	mins/ person/ day	loads/ person /day	Lpd/ person [gpd/ person]	*Formula*[ξξ]
23	Toilet 1	Dual flush	4.9 [1.3]			2.55[YY]			12.5 [3.3]	*=C23*F23*
24	Toilet 2	Dual flush	4.9 [1.3]			2.55[YY]			12.5 [3.3]	*=C24*F24*
25										
26										
27										
28	Laundry	Front Loading, High-efficiency			76 [20.1]			0.37	28.1 [7.4]	*=E28*H28*
29										
30						**Total of All Fixtures**			53 [14]	*=sum (I23:I29)*
31						**Number of Occupants**			4	*=I11*[Y]
32						**Total Daily Indoor Rainwater Demand, Lpd [gpd]**			212 [56]	*=I30*I31*
33						**Percentage of Indoor Water to Be Supplied with Rainwater,%**			0.45	*=I32/I12*[Y]
34						**Monthly Indoor Rainwater Demand, L/month [gal/month]**			6,445 [1,702]	*=I33/I13*[Y]

τ -Taken from actual manufacturer's rating or from Table 3.2.
ττ - Taken from actual water audit or from Table 3.2.
Y - From Step 1.3.

YY - Lpd/person from Table 3.2 divided by 2 to account for 2 toilets in the home.
ξξ - The formulas contained within the cells are shown, but they would not be visible as such in your spreadsheet

**STEP 1.5
OUTDOOR RAINWATER
DEMAND**

Step 1.5: Outdoor Rainwater Demand

When it comes to estimating outdoor volumes, the reality is that we would never actually design outdoor rainwater landscape irrigation without first considering how it fits within the larger context of full-property water security (see Chapter 1).

Also, because actual landscape irrigation volumes are so heavily influenced by site-specific factors such as types of vegetation, annual vs perennial plants, mulching, soil carbon, and wind, as well as the climate influences of rainfall, temperature, and humidity, there's no one easy way to approach determining/estimating the required irrigation volumes for a property.

As such, if you are looking for a good determination of landscape irrigation volumes, we recommend that you locate your nearest irrigation specialist and talk to him/her. Look for someone who has experience with your type of landscape, be it a garden, a greenhouse, or even just a lawn. Someone who's already dealt with a local landscape will have a much better idea of the typical volumes you'll need for your particular situation and may even be able to provide some very useful rules of thumb.

If you require more precise irrigation estimates, the USDA's National Engineering Handbook *Irrigation Guide* is available free online (listed in the Resources section). It contains methods and tables for estimating irrigation based on soil type, crop types, infiltration, ground water, application efficiency, and the design and planning of numerous types of irrigation systems.

For a rough estimate of what your irrigation requirements might be, we suggest following the guidance provided in *Rainwater Harvesting for Drylands and Beyond,* or this quick-and-dirty, simplified approach adapted from *The New Create an Oasis With Greywater* by Art Ludwig.

You'll need to get following data:

- Average midsummer high. Take the highest number found for the daily maximum temperature from your Climate Normals. Said another way, this is the average daily high for the hottest month.
- Average midsummer relative humidity. The highest relative humidity in the summer will determine if you are in a dry climate (less than 50%) or if you are in a humid climate (more than 50%).
- The area you plan to irrigate in m² (ft²).

Using the legend in Table 3.3, determine which classification best describes your climate regime: cool humid, cool dry, warm humid, warm dry, hot humid, or hot dry, and find your approximate ranges for Average Peak Evapotranspiration and Estimated Irrigation, both of which are provided as ranges.

Table 3.3

Classification	Average Peak Evapotranspiration mm/month [in/month]	Estimated Irrigation Liter/m² per month [gal/ft² per month]
Cool Dry	108 – 151 [4.2 – 5.9]	108 – 151 [2.6 – 3.7]
Cool Humid	77 – 108 [3 – 4.2]	77 – 108 [1.9 – 2.6]
Hot Dry	215 – 348 [8.5 – 13.7]	215 – 348 [5.3 – 8.5]
Hot Humid	151 – 215 [5.9 – 8.5]	151 – 215 [3.7 – 5.3]
Warm Dry	151 – 198 [5.9 – 7.8]	151 – 198 [3.7 – 4.9]
Warm Humid	108 – 151 [4.2 – 5.9]	108 – 151 [2.6 – 3.7]

Cool = under 21°C [70°F] average midsummer high
Warm = 21°C to 32°C [70°F to 90°F] average midsummer high
Hot = over 32°C [90°F] average midsummer high
Humid = over 50% average midsummer relative humidity
Dry = under 50% average midsummer relative humidity

Table 3.3: *Peak evapotranspiration and estimated irrigation values by climate.*

SOURCE: ADAPTED FROM: LUDWIG, A. 2015. THE NEW CREATE AN OASIS WITH GREYWATER: CHOOSING, BUILDING, AND USING GREYWATER SYSTEMS, TABLE 2.5. OASIS DESIGN, SANTA BARBARA, CA

If the average peak evapotranspiration for your hottest month is greater than your monthly rainfall in that same month, then soil will dry out unless irrigation is provided.

Your rough estimate for the amount of irrigation required is determined by multiplying the Estimated Irrigation number, in liters/m² per month (gal/ft² per month) by the area you wish to irrigate.

Lastly, you'll want to state what percentage of the total required irrigation that you intend on supplying with rainwater and which months you intend on irrigating. If you intend right from the onset on supplying 100% of your irrigation with rainwater, your Outdoor Rainwater Demand is simply equal to 100% of your Total Estimated Irrigation for the months you intend on irrigating.

Step 1.5 Template

If you intend on using the methodology above for estimating your outdoor irrigation requirements, follow the Step 1.5 template.

If you use a different methodology for estimating your irrigation, simply ensure that you state your Total Estimated Irrigation in cell B6, your percentage of irrigation to supply with rainwater in cell B7, and the subsequent outdoor rainwater demand in cell B8. That way, the reference cells in the subsequent steps will still work.

Step 1.5 Template: Outdoor Rainwater Demand

	A	B	C
1	Garden Area, m² [ft²]		
2			
3		Min	Max
4	Estimated Irrigation$^\tau$, Liter/m² per month [gal/ft² per month]		
5	Estimated Irrigation, Liters/month [gal/month]	=B4*B1	=C4*B1
6	Total Estimated Irrigation, Rounded$^{\tau\tau}$, Liters/month [gal/month]		
7	Percentage of Irrigation to Be Supplied with Rainwater^Y, %		
8	Outdoor Rainwater Demand, Liters/month [gal/month]	=B6*B7	

τ -*From Table 3.3, based on the appropriate classification, input the minimum and the maximum estimated values in column B and C*
$\tau\tau$ -*Select a value within the min and max range. Round this number to an appropriate number of significant digits.*
Y - *If you plan to irrigate 100% with harvested rainwater, input 100%.*

Example: Outdoor rainwater demand and estimating irrigation (Step 1.5)

Continuing on from the previous example: The family is also planning to irrigate their 100 m² (1,076 ft²) backyard garden with rainwater.

Vancouver, Canada, is classified as Warm Humid in Table 3.3 because the midsummer high is between 21–32°C (70–90°F), and the average midsummer humidity is above 50%.

From Step 1.1, the rainfall in the hottest month (August) is 39.1 mm (1.5 in). From Table 3.3 the average peak evapotranspiration is 108–151 mm/month (4.2–5.9 in/month). Therefore, unless they provide irrigation, the soil will dry out.

Based on the calculation below, the Total Estimated Irrigation that their garden will need for August is somewhere between 10,800–15,100 liters/month (2,798–3,981 gal/month). They realize that this number is only ballpark, so they decide to round their irrigation requirement to 15,000 liters/month (3,963 gal/month).

They intend on attempting to supply 100% of their garden irrigation with rainwater; therefore, they set their Outdoor Rainwater Demand to 15,000 liters/month (3,963 gal/month) for each month they plan on irrigating: June, July, August, and September.

Step 1.5 Example

	A	B	C
1	Garden Area, m² [ft²]	100 [1,076]	
2			
3		Min	Max
4	Estimated Irrigation[τ], Liter/m² per month [gal/ft² per month]	108 [2.6]	151 [3.7]
5	Estimated Irrigation, Liters/month [gal/month]	10,800 [2,798]	15,100 [3,981]
6	Total Estimated Irrigation, Rounded[ττ], Liters/month [gal/month]	15,000 [3,963]	
7	Percentage of Irrigation to Be Supplied with Rainwater[Y], %	100%	
8	Outdoor Rainwater Demand, Liters/month [gal/month]	15,000 [3,963]	

τ -From Table 3.3, based on the appropriate classification, input the minimum and the maximum estimated values in column B and C
ττ -Select a value within the min and max range. Round this number to an appropriate number of significant digits.
Y -For meeting 100% of irrigation needs with harvested rainwater, input 100%.
Calculations in each choice of unit were performed independently and values in earlier input fields rounded to a significant number of digits. The end result is a rounding discrepancy if you compare the metric vs US Customary results.

Measuring Actual Demand

If you are looking to provide rainwater for an existing home, there's another great way to determine both indoor and outdoor demand: Look at your past water utility bills.

A past water utility bill will show you the total volume of water used per month in your home, but your bill will include both indoor and outdoor as one total amount. If you are planning on replacing your entire water demand, you could take the monthly volumes and find averages for each month (use several years of information, if possible).

There is also a way to estimate the split between indoor (domestic) and outdoor uses, but it only works if you are in a climate where there are multiple months per year where you do not irrigate. An example of this would be in a cold climate where you only garden from May through to September. If that's the case, the months that you do not irrigate (November to April) should be fairly representative of your indoor water demand. You use these months (again, using amounts over several years) to come up with an average Indoor Rainwater Demand volume.

For the months that you are irrigating, you'll want to look and see if there is a month with a spike in volume usage. Again, it's best to average each month over several years. Subtract the highest average month by your Indoor Rainwater Demand, and you'll have an idea of what your peak Outdoor Rainwater Demand might be.

Step 1.6 Total Rainwater Demand

Now, you must combine the Indoor Rainwater Demand and the Outdoor Rainwater Demand and build a table on a month-by-month basis.

Unless you have some special considerations, such as a seasonal dwelling, or known seasonal increases in occupants, your indoor volumes can be assumed to stay constant for each month of the year.

Your Outdoor Rainwater Demand volumes are only for those months that you intend on irrigating, which for many places in North America, is only in the spring and summertime.

To be conservative, we typically recommend that you take the peak irrigation volume from Step 1.5 and use this constant volume for all irrigation months. However, if you've used a different method to estimate irrigation, you may have varying volumes throughout your irrigation season.

In this template, you simply pull the Indoor Rainwater Demand (Step 1.4 cell I34) into column B and the Outdoor Rainwater Demand (Step 1.5 cell B8) into column C to arrive at your Total Rainwater Demand (column D), on a month-by-month basis and in total liters/year (gal/year).

In the last row of the template (row 16), the Indoor Rainwater Demand and the Outdoor Rainwater Demand are stated as a percentage of the Total Rainwater Demand.

Step 1.6 Template: Total Rainwater Demand

	A	B	C	D
1	Month	Indoor Rainwater Demand$^\tau$	Outdoor Rainwater Demand$^{\tau\tau,Y}$	Total Rainwater Demand
2		liters [gal]	liters [gal]	liters [gal]
3	Jan	= Step 1.4 cell I34		=B3+C3
4	Feb	= Step 1.4 cell I34		=B4+C4
5	Mar	= Step 1.4 cell I34		=B5+C5
6	Apr	= Step 1.4 cell I34		=B6+C6
7	May	= Step 1.4 cell I34		=B7+C7
8	Jun	= Step 1.4 cell I34	= Step 1.5 cell B8	=B8+C8
9	Jul	= Step 1.4 cell I34	= Step 1.5 cell B8	=B9+C9
10	Aug	= Step 1.4 cell I34	= Step 1.5 cell B8	=B10+C10
11	Sep	= Step 1.4 cell I34	= Step 1.5 cell B8	=B11+C11
12	Oct	= Step 1.4 cell I34		=B12+C12
13	Nov	= Step 1.4 cell I34		=B13+C13
14	Dec	= Step 1.4 cell I34		=B14+C14
15	Total Rainwater Demand, liters/yr [gal/yr]	=SUM(B3:B14)	=SUM(C3:C14)	=SUM(D3:D14)
16	Percentage of Total Rainwater Demand, %	=B15/D15	=C15/D15	

τ - *Domestic water usage assumed to be constant throughout the year (i.e. over 12 months)*

$\tau\tau$ - *Only for the months that irrigation is planned to be used. In this example it is assumed June — September*

Y - Consider using a more precise alternative method for estimating irrigation volumes depending on your needs and context

Example: Total rainwater demand (Step 1.6)

To calculate the Total Rainwater Demand, the Vancouver family starts another tab in their spreadsheet and builds a table by following the Step 1.6 Template:

Their Total Average Rainwater Demand is shown on a month-by-month basis and results in a total demand of 137,340 liters/year (36,276 gal/yr).

Of the annual rainwater they intend on harvesting, 56% is currently planned for indoor use and 44% is planned for outdoor use.

Refer to the Step 1.6 Template to see the formulas in each cell.

Step 1.6 Example: Total Rainwater Demand

	A	B	C	D
	Month	Indoor Rainwater Demand$^\tau$	Outdoor Rainwater Demand$^{\tau\tau,Y}$	Total Rainwater Demand
2		liters [gal]	liters [gal]	liters [gal]
3	Jan	6,445 [1,702]	0 [0]	6,445 [1,702]
4	Feb	6,445 [1,702]	0 [0]	6,445 [1,702]
5	Mar	6,445 [1,702]	0 [0]	6,445 [1,702]
6	Apr	6,445 [1,702]	0 [0]	6,445 [1,702]
7	May	6,445 [1,702]	0 [0]	6,445 [1,702]
8	Jun	6,445 [1,702]	15,000 [3,963]	21,445 [5665]
9	Jul	6,445 [1,702]	15,000 [3,963]	21,445 [5665]
10	Aug	6,445 [1,702]	15,000 [3,963]	21,445 [5665]
11	Sep	6,445 [1,702]	15,000 [3,963]	21,445 [5665]
12	Oct	6,445 [1,702]	0 [0]	6,445 [1,702]
13	Nov	6,445 [1,702]	0 [0]	6,445 [1,702]
14	Dec	6,445 [1,702]	0 [0]	6,445 [1,702]
15	Total Rainwater Demand, liters/yr [gal/yr]	77,340 [20,424]	60,000 [15,852]	137,340 [36,276]
16	Percentage of Total Rainwater Demand, %	0.56	0.44	

τ - Domestic water usage assumed to be constant throughout the year (i.e. over 12 months)
$\tau\tau$ - Only for the months that irrigation is planned to be used. In this example it is assumed June — September
Y - Consider using a more precise alternative method for estimating irrigation volumes depending on your needs and context
Calculations in each choice of unit were performed independently and values in earlier input fields rounded to a significant number of digits. The end result is a rounding discrepancy if you compare the metric vs US Customary results.

Step 1.7: Supply-to-Demand Ratio

In this exercise we are looking at the amount of rainwater you have harvested and comparing that against the amount of rainwater you intend to use. It could be called the Rainwater Supply-to-Rainwater Demand Ratio, but we simply call it Supply-to-Demand Ratio for simplicity's sake. The differentiation is important however, because you could also calculate Rainwater Supply to Total Demand, where Total Demand would include the water volumes you intend on meeting with your municipal system, for instance.

The Supply-to-Demand Ratio is calculated as follows:

Supply-to-Demand Ratio = Total Average Rainwater Supply, liters/yr (gal/yr) ÷ Total Rainwater Demand, liters/yr (gal/yr)

In Step 1.7, you simply take the Total Average Supply (Step 1.2 cell C18) and divide that number by the Total Rainwater Demand (Step 1.6 cell D15). Then you ask:

If your ratio is below 1.0, then you are trying to use more rainwater than you can actually harvest on an annual basis.

You need to go back and either reduce your rainwater demand, or increase your supply, or do

combination of both. Here are the precise ways and options to do this:

- Go back to Step 1.2 and increase your roof catchment in cell B1.
- Go back to Step 1.3 and reduce your total indoor water demand by changing out fixtures, removing fixtures, or changing the usage patterns (columns C-H).
- Go back to Step 1.4 and reduce your indoor rainwater demand by changing out rainwater fixtures, removing rainwater fixtures (columns C-H) or simply reduce the percentage of indoor water that you intend to supply with rainwater in cell I33.
- Go back to Step 1.5 and reduce the percentage of irrigation that you intend on supplying with rainwater (B7). Alternatively, consider a less conservative approach to estimating irrigation volumes, or reduce your irrigation requirements using the suggestions in Chapter 1.

The Supply-to-Demand Ratio is about making sure that your planned uses for rainwater are not more than what your "rainfall budget" will allow. This can often be a reality check; the average indoor and outdoor water consumption rates in North American homes vastly outstrip the amount of rain that could be harvested by those same homes.

To give you an idea of this, let's look at some hypothetical situations in Table 3.4. We've chosen three North American cities with different annual rainfall amounts. Assuming that the homeowners in each of these cities were hoping to replace their indoor water consumption with 100% rainwater, they'd end up with the supply-to-demand ratios indicated in column G.

Step 1.7 Template: Supply-to-Demand Ratio

	A	B
20	Total Average Supply, liters/yr [gal/yr]	*=Step 1.2 cell C18*
21	Total Rainwater Demand, liters/yr [gal/yr]	*=Step 1.6 cell D15*
22	Supply / Demand Ratio	*=B20/B21*

In each scenario, the Supply-to-Demand Ratio is well below 1.0. Assuming that they cannot increase their catchment area, they can either substantially reduce their Total Water Demand (column D) or reduce their expectations of how much demand will be met by rainwater (column E). They must change one or both until the Supply-to-Demand Ratio (column G) is at, or above, 1.0.

Now, although you cannot proceed to Part 2 until your Supply-to-Demand Ratio is above 1.0, we must warn you that having a Supply-to-Demand Ratio above 1.0 does not guarantee that you have a viable system. You'll still need to complete Part 2 and Part 3 of the feasibility stage.

It's actually a good idea to aim for as high of a supply-to-demand ratio as reasonable at this point, and especially so if you are in a climate with uneven rain distribution patterns or if you are designing an off-grid system with no backup or where backup is expensive.

Note, lastly, that the monthly time step assumption discussed at the beginning of this chapter will *underestimate* supply in the spreadsheet model. If you are slightly below 1.0 and your design is constrained such that you can't reduce demand or increase supply, you may consider evaluating your supply-to-demand ratio on a daily basis. The increased granularity (daily vs monthly) may provide a less-conservative result and increase your assurance of system viability.

Table 3.4:

Calculating the Supply-to-Demand Ratio based on 100% rainwater supply and average North American indoor water usage for three different cities.

Table 3.4: Rainfall Budget

A	B	C	D	E	F	G
	Annual Rainfall	Average Supply[τ]	Total Water Demand[ττ]	Rainwater Demand	Rainwater Demand[Y]	Supply to Demand Ratio[YY]
City, State	mm [in]	liters/yr [gal/yr]	liters/yr [gal/yr]	%	liters/yr [gal/yr]	
Vancouver, BC	1,158 [45.6]	260,550 [68,830]	386,900 [102,200]	100%	386,900 [102,200]	0.67
Flagstaff, AZ	581 [22.9]	130,725 [34,553]	386,900 [102,200]	100%	386,900 [102,200]	0.34
Worchester, MA	1,248 [49.1]	280,800 [74,084]	386,900 [102,200]	100%	386,900 [102,200]	0.73

τ - Based on average North American house size (catchment area) of 250 m² [2691 ft²] and 0.9 roof efficiency.
ττ - Based on average North America per person indoor water volume of 265 liters/day [70 gal/day] and four people per household.

Y - Rainwater Demand (column E) * Total Water Demand (column D)
YY - Average Supply (column C) / Rainwater Demand (column F)

EXAMPLE

Example: Calculating the Supply-to-Demand ratio (Step 1.7)

In their spreadsheet, directly below their Step 1.6 table, the Vancouver homeowners build the Step 1.7 table by following the template provided.

They reference the Total Average Supply from Step 1.2 cell C18 and the Total Rainwater Demand from Step 1.6 cell D15. Then they divide the former by the latter.

The ratio is greater than 1.0, so they continue on to Part 2, Optimize Your Storage.

Step 1.7 Example

	A	B
20	Total Avg Supply, liters/yr [gal/yr]	219,393 [57,979]
21	Total Avg Demand, liters/yr [gal/yr]	137,340 [36,276]
22	Supply / Demand Ratio	1.6

Part 2. Optimize Your Storage

Balancing supply and demand is really just a fancy way of saying: *Ensure that you don't run out of rainwater for any given month.* This means that you never have a zero volume in your tank.

However, the goal is not only to balance your system, it's to balance your system *for the least possible cost.* This is what we call *optimized.* Optimized storage performance is the goal/output that we are looking for throughout Part 2.

Assuming that your collection area (i.e. roof size) is fixed, there are two different directions from which you can approach optimization:

- Approach (i): Establish your rainwater demands and then back-solve your storage capacity to optimize your storage performance.
- Approach (ii): Establish your storage capacity and then back-solve your demand to optimize your storage performance.

What is back-solving? Also called *goal-seeking* or a *what-if analysis,* it's simply an iterative process in which you are seeking an input for a given output.

Approach (i) is ideal when you have no space constraints and/or no significant budget constraints, or when your rainfall budget is large and/or your demands are small relative to your rainfall budget. But that's not the reality for most people most of the time.

Approach (ii) works if you're quite happy "getting whatever you get." We see this approach used most often in supplemental outdoor rainwater-supply scenarios. Consider your neighbor who simply put in a small rain barrel beside her garden. She purchased a tank that fit her budget and will supplement with rainwater when it is there.

However, if you are actually interested in designing and planning a system and your primary goal is *best storage performance for a particular budget,* you'll need a combination of both approaches.

Regardless of how you want to approach optimizing your system, the next thing you have to do is set up your spreadsheet tool by following the template for Part 2.

Lastly, we use the words "storage capacity" instead of "tank size" because at this point you haven't decided on your tank configuration or even how many tanks you will install. Storage capacity represents the total volume of all of the tanks you intend on installing. In small systems, you can interchangeably use the words *storage capacity* and *rain tank size.*

The Part 2 Template

In Part 2, we are performing a storage calculation on a month-by-month basis. This is the trickiest part of sizing a RWH system.

Here are the things you are trying to consider:

- You need to look at monthly supply versus demand over multiple years because it can take several years for a system to find equilibrium. We recommend that you look at the start-up year (Year 0) plus four more years, for a total of five years.

- Some months may have a positive net supply (more supply than demand) and other months may have a negative net supply (less supply than demand). At the end of every month, you want to establish the stored volume. The volume at the end of the current month is simply the volume at the end of the prior month plus the net supply for the current month. If the net supply is positive, your stored volume goes up, if the net supply is negative, your stored volume goes down.

- You can't have a lower storage volume than zero, and you can't have a higher storage volume than your tank capacity; any volumes beyond your tank capacity are simply overflowed.

Part 2 Template: Optimize Your Storage

	A	B	C	D	E	F	G	H	I	J
1	Roof Catchment, m^2 [ft^2]			=Step 1.2 cell B1						
2	Roof Efficiency			=Step 1.2 cell B2						
3	Storage Capacity, liters [gal]									
4	Commission Month			Mar						
5	Initial Tank Charge, liters [gal]									
6	Supply-to-Demand Ratio			=Step 1.7 cell B22						
7							Storage Performance			
8	Month	Average Rainfall	Supply	Demand	Net Supply$^\tau$	Year 0$^{\tau\tau,Y,YY}$	Year 1$^{\tau\tau,YY}$	Year 2$^{\tau\tau}$	Year 3$^{\tau\tau,\psi}$	Year 4$^{\tau\tau,\psi}$
9		Depth	Volume	Volume	Volume	Month End Volume	Month End Volume	Month End Volume	Month End Volume	Month End Volume
10	Jan	=Step 1.2 cell B6	=Step 1.2 cell C6	=Step 1.6 cell D3	=C10-D10		=IF(F21+E10<0,0, IF(F21+E10>D3,D3, F21+E10))$^\xi$	=IF(G21+E10<0,0, IF(G21+E10>D3,D3, G21+E10))$^\xi$	ξ	ξ
11	Feb	=Step 1.2 cell B7	=Step 1.2 cell C7	=Step 1.6 cell D4	=C11-D11		=IF(G10+E11<0,0, IF(G10+E11>D3,D3, G10+E11))^{YY}	=IF(H10+E11<0,0, IF(H10+E11>D3,D3, H10+E11))^{YY}		
12	Mar	=Step 1.2 cell B8	=Step 1.2 cell C8	=Step 1.6 cell D5	=C12-D12	=IF(D5+E12<0,0, IF(D5+E12>D3,D3, D5+E12))^Y	=IF(G11+E12<0,0,I F(G11+E12>D3,D3, G11+E12))	=IF(H11+E12<0,0, IF(H11+E12>D3,D3, H11+E12))		
13	Apr	=Step 1.2 cell B9	=Step 1.2 cell C9	=Step 1.6 cell D6	=C13-D13	=IF(F12+E13<0,0, IF(F12+E13>D3,D3, F12+E13))^{YY}	=IF(G12+E13<0,0, IF(G12+E13>D3,D3, G12+E13))	=IF(H12+E13<0,0, IF(H12+E13>D3,D3, H12+E13))		
14	May	=Step 1.2 cell B10	=Step 1.2 cell C10	=Step 1.6 cell D7	=C14-D14	=IF(F13+E14<0,0, IF(F13+E14>D3,D3, F13+E14))	=IF(G13+E14<0,0,I F(G13+E14>D3,D3, G13+E14))	=IF(H13+E14<0,0, IF(H13+E14>D3,D3, H13+E14))		
15	Jun	=Step 1.2 cell B11	=Step 1.2 cell C11	=Step 1.6 cell D8	=C15-D15	=IF(F14+E15<0,0, IF(F14+E15>D3,D3, F14+E15))	=IF(G14+E15<0,0, IF(G14+E15>D3,D3, G14+E15))	=IF(H14+E15<0,0, IF(H14+E15>D3,D3, H14+E15))		
16	Jul	=Step 1.2 cell B12	=Step 1.2 cell C12	=Step 1.6 cell D9	=C16-D16	=IF(F15+E16<0,0, IF(F15+E16>D3,D3, F15+E16))	=IF(G15+E16<0,0, IF(G15+E16>D3,D3, G15+E16))	=IF(H15+E16<0,0, IF(H15+E16>D3,D3, H15+E16))		
17	Aug	=Step 1.2 cell B13	=Step 1.2 cell C13	=Step 1.6 cell D10	=C17-D17	=IF(F16+E17<0,0, IF(F16+E17>D3,D3, F16+E17))	=IF(G16+E17<0,0, IF(G16+E17>D3,D3, G16+E17))	=IF(H16+E17<0,0, IF(H16+E17>D3,D3, H16+E17))		
18	Sep	=Step 1.2 cell B14	=Step 1.2 cell C14	=Step 1.6 cell D11	=C18-D18	=IF(F17+E18<0,0, IF(F17+E18>D3,D3, F17+E18))	=IF(G17+E18<0,0,I F(G17+E18>D3,D3, G17+E18))	=IF(H17+E18<0,0, IF(H17+E18>D3,D3, H17+E18))		
19	Oct	=Step 1.2 cell B15	=Step 1.2 cell C15	=Step 1.6 cell D12	=C19-D19	=IF(F18+E19<0,0, IF(F18+E19>D3,D3, F18+E19))	=IF(G18+E19<0,0, IF(G18+E19>D3,D3, G18+E19))	=IF(H18+E19<0,0, IF(H18+E19>D3,D3, H18+E19))		
20	Nov	=Step 1.2 cell B16	=Step 1.2 cell C16	=Step 1.6 cell D13	=C20-D20	=IF(F19+E20<0,0, IF(F19+E20>D3,D3, F19+E20))	=IF(G19+E20<0,0, IF(G19+E20>D3,D3, G19+E20))	=IF(H19+E20<0,0, IF(H19+E20>D3,D3, H19+E20))		
21	Dec	=Step 1.2 cell B17	=Step 1.2 cell C17	=Step 1.6 cell D14	=C21-D21	=IF(F20+E21<0,0, IF(F20+E21>D3,D3, F20+E21))	=IF(G20+E21<0,0, IF(G20+E21>D3,D3, G20+E21))	=IF(H20+E21<0,0,I F(H20+E21>D3,D3, H20+E21))		

τ - Net Supply is the supply volume (column C) minus the demand column (column D)

$\tau\tau$ - Ensure that the month end volume does not go below 0, or above the storage capacity (D3). The formulas provided show how to do this using an IF function.

Y - Start in Year 0 (column F) and the row associated with the commission month (D4). Add the initial tank charge (D5) to the net supply for that month.

YY - For the next month, add the prior month end volume to the current month net supply. Continue this pattern until the end of the year.

ξ - At the start of all subsequent years (column G, H, I or J), add the prior December month end volume (row 21) to the net supply for January (E10).

Ψ - Continue on with the same formula patterns in year 3 (column I) and year 4 (column J). The equations are not shown.

The first thing you'll notice in the Part 2 Template is that columns B–D simply pull data and calculations from all of the previous steps. The Net Supply (column E) is the monthly supply minus the monthly demand.

Then there's columns F, G, H, I, and J. These columns represent your storage performance, and the cells within give you the volume of water stored in your tank(s) at the end of the month. These are the columns that you'll be watching closely as you back-solve other values in an attempt to optimize your system.

Column F is your start-up year, Year 0. In the Part 2 Template, we show that the commission month is March (as indicated in cell D4); therefore, the commission month formula is in the row associated with March (cell F12). Note that you'll want to set up your table with the commission month equation in the correct row, which very well may *not* be March for your particular scenario.

The second month equation is shown in Year 0, April (cell F13).

Now, don't be too intimidated by the IF functions. Look closely at the formula in this cell and you'll see it's pretty straightforward:

Basically, we are simply using IF functions to ensure that the numbers in our spreadsheet never go negative and never exceed the specified storage capacity. The equation pattern repeats itself all the way until December (row 21). At the start of the next year, Year 1 (cell G10), a slightly different pattern occurs to account for the relative positioning of December vs January. The principle is the same though — we are simply adding together the prior month tank volume to the current month net supply. From there you continue so on and so forth for columns H, I, and J.

Take your time following this template and setting up your own Part 2 spreadsheet table to ensure that the formulas are correct in each row and in each column. And although it may seem time consuming to set up, you'll see in a moment why using it will pay immense time dividends.

`=IF(F12+E13<0,0,    IF(F12+E13>D3,D3,    F12+E13))`

| if the tank volume at the end of March + the net supply in April is less than 0, use 0 | if the tank volume at the end of March + the net supply in April is more than the tank capacity, then use tank capacity | Otherwise, sum together the volume at the end of March + the net supply in April |

The Part 2 Process

The step-by-step process for the second part of the feasibility stage is shown in Figure 3.3.

Once again, notice the cyclical, or iterative, nature of this process. Until you have an acceptable design, you'll go 'round and 'round — adjusting catchment area, storage volume, or your demands.

Each step is described in detail in the following sections.

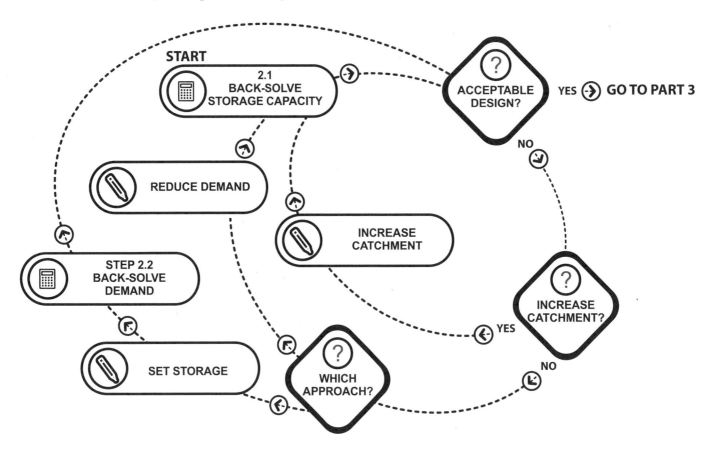

Fig. 3.3: *The step-by-step process for Part 2 of the feasibility stage: Optimize Your Storage.*

Credit: Verge Permaculture/ S. Fidler

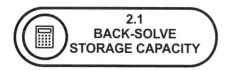

2.1
BACK-SOLVE
STORAGE CAPACITY

Step 2.1: Back-Solve Storage Capacity

Since you started this feasibility process with defining your supply and demand volumes in Part 1, it makes sense to start Part 2 with back-solving the storage capacity that optimizes the demand volumes you've already defined (Approach i).

Mathematically speaking, supply and demand is balanced when there are no zeros in the storage performance columns (G, H, I, J) for the Month-end Volumes. Table 3.5 shows you a hypothetical system that is not balanced, because there are zeros in August and September throughout years 1–4.

Remember also that the goal is not only to balance a system, it's to optimize it, which is to balance a system for the least possible cost.

Here's the step-by-step methodology to back-solve for optimized storage capacity:

i. Estimate a decent starting point for your first iteration of storage capacity. For off-grid systems, an okay rule of thumb is to take your largest three demand months and add them together. RWH systems that have back-up from another source can use substantially smaller starting tank sizes.

ii. Input that storage capacity into cell D3 in your Part 2 table and let your spreadsheet do the math for you!

iii. Look closely at the storage performance columns (G, H, I, J). Are there any months with a zero month-end volume?
   - No: Your tank may be sufficient or over-sized. Skip to (iv).
   - Yes: Your tank is too small. Add more capacity in small increments (i.e. put a larger number in cell D3) until the smallest number in the storage performance columns barely exceeds zero. Skip to (v).

iv. Lower your storage capacity (cell D3) in small increments. When you hit a zero volume in the storage performance columns, back off and go back up again. Add in even smaller increments of storage capacity until the smallest number barely exceeds zero.

v. You've now found the absolute minimum storage capacity that will balance your

Table 3.5: An example of supply and demand not balanced

	A	F	G	H	I	J
7			Storage Performance			
8	Month	Year 0	Year 1	Year 2	Year 3	Year 4
9		liters [gal]	liters [gal]	liters [gal]	liters [gal]	liters [gal]
10	Jan	0 [0]	2,500 [661]	2,500 [661]	2,500 [661]	2,500 [661]
11	Feb	0 [0]	2,500 [661]	2,500 [661]	2,500 [661]	2,500 [661]
12	Mar	0 [661]	2,500 [661]	2,500 [661]	2,500 [661]	2,500 [661]
13	Apr	0 [661]	2,500 [661]	2,500 [661]	2,500 [661]	2,500 [661]
14	May	2,500 [661]	2,500 [661]	2,500 [661]	2,500 [661]	2,500 [661]
15	Jun	1,397 [374]	1,397 [374]	1,397 [374]	1,397 [374]	1,397 [374]
16	Jul	6 [12]	5 [12]	5 [12]	5 [12]	5 [12]
17	Aug	0 [0]	0 [0]	0 [0]	0 [0]	0 [0]
18	Sep	0 [0]	0 [0]	0 [0]	0 [0]	0 [0]
19	Oct	1,493 [391]	1,493 [391]	1,493 [391]	1,493 [391]	1,493 [391]
20	Nov	2,500 [661]	2,500 [661]	2,500 [661]	2,500 [661]	2,500 [661]
21	Dec	2,500 [661]	2,500 [661]	2,500 [661]	2,500 [661]	2,500 [661]

supply and demand. Is this optimized? Not quite. Here are some reasons why:

○ A natural sludge layer will form at the bottom of your tank. You need to maintain at least 100 mm (4 inches) of water in your tank so that you do not disturb this sludge layer.

○ It's a far better and more robust design to make sure that you have at least one month of demand volume stored away. Think of this as your peace-of-mind water. Depending on your supply scenario and risk tolerance, you may want to specify a larger or smaller minimum stored volume.

○ Are you wanting to hold back some water for emergency scenarios, such as fire protection? Here you would specify that volume allowance.

○ And of course, the sizing above is entirely dependent on you receiving the average amount of rain. You'll want to be thinking about this, even though in Part 3 we'll be ensuring our design is robust enough for low-rainfall scenarios.

After you make the above considerations, set your minimum month-end volume in your mind, and go back to your storage capacity (cell D3). Slowly increase the storage capacity until the smallest number in the storage performance columns (G, H, I, J) exceeds that minimum volume.

You have now found the storage capacity that, for the roof size, demands, and average rainfall that you have defined, results in optimized storage performance.

Acceptable Design?

Now you merely consider if your combination of roof catchment, storage capacity, and rainwater demand volumes are acceptable. It's a question of whether your design meets your needs for your budget.

The cost of storage can be highly variable depending on your location and the tank material and type selected, so it's advised that you seek out local resources and information when it comes to getting reliable costing information. But for back-of-a-napkin costing you can probably use 25–50 cents per liter, or 1–2 USD per gallon.

If you've just back-solved for Storage Capacity (Step 2.1):

If your Net Supply volume (column E) is mostly a positive number, with perhaps only a few months dropping below zero, you likely will have found quite a reasonable tank size. Hurray, you get to move on to Part 3!

However, if your monthly demands and supply volumes are substantially mismatched, you'll quickly find yourself with an incredibly unrealistic, very large, and very expensive optimized storage capacity. Back to the drawing board it is.

If you've just back-solved for Demands (Step 2.2):

If you are arriving from Step 2.2 (meaning that you've already been around the loop at least once), you are now considering if the demands that you've just back-solved will meet your water needs. If yes, move on to Part 3. If no, continue around once again.

Increase Catchment?

If it's possible, your first choice should be to try to increase your catchment (such as adding the roof of a garage or outbuilding to your collection area). If you can increase catchment (or would at least consider it), go back to Step 1.2 cell B1 and enter a larger roof area. Then return to the Part 2 table and repeat the instructions for Step 2.1, back-solving once again the optimized tank capacity.

If you can't increase catchment, follow the "No" path in Figure 3.3.

Which Approach?

You now have choice of two different approaches:

1. Reduce your demands (Steps 1.3–1.7), then back-solve for storage capacity again (Step 2.1).

    OR

2. Set the storage capacity your budget (or space) will allow, then back-solve for demands (Step 2.2).

Regardless of which approach you choose, you don't want to iterate blindly, which could easily result in a waste of time and no possible solution. It's well worth looking closely at the numbers in the Part 2 table to first assess what changes in demand will most likely result in a balanced system.

Ask yourself the following questions:

- In which months does my storage go to zero?
- In which months are my net supply volumes most mismatched (i.e. the largest negative numbers in column E)?

From there: *Is there a way I can reduce demands to better match supply and demand on a monthly basis?* Then, continue on with one of the two possible approaches.

Reduce Demand

If you go this route, you'll want to do some, or all, of the following:

- Go back to Step 1.3 and reduce your total indoor water demand by changing out fixtures, removing fixtures, or changing the usage patterns (columns C–H).
- Go back to Step 1.4 and reduce your indoor rainwater demand by changing out rainwater fixtures, removing rainwater fixtures (columns C–H), or by simply reducing the percentage of indoor water that you intend to supply with rainwater (cell I33).
- Go back to Step 1.5 and reduce the percentage of irrigation that you intend on supplying with rainwater (cell B7). Alternatively, consider a less conservative approach to estimating irrigation volumes.

Once you've made a change (or several changes), go back to your table for Part 2 and follow the instructions again for Step 2.1.

Set Storage and Step 2.2 Back-Solve Demand

If you elect this approach, you'll now be setting your maximum storage capacity to whatever it is your budget or space will allow. Then you'll attempt to establish the demand volume(s) that ensure your system is optimized for the supply you expect.

The process is exactly the opposite of what we were doing in Step 2.1. Here is the step-by-step methodology to back-solve for demand volumes:

i. Input your maximum storage capacity into Part 2 table, cell D3.

ii. Are there any months with a zero month-end volume in the storage performance columns (G, H, I, J)?

     ○ No: Your demands are sufficient or under-sized. Skip to (iii).

     ○ Yes: Your demands are too big. Reduce your demand(s) in small increments in Steps 1.3, 1.4, and/or 1.5 until the smallest number in the storage performance columns barely exceeds zero. Skip over (iii).

iii. Increase your demands in small increments in Steps 1.3, 1.4, and/or 1.5 until you hit a zero volume in the storage performance columns. Then back off slightly and reduce by an even smaller increment of demand until the smallest number barely exceeds zero.

Note that when back-solving demands for a set storage capacity, our preferred method

is to iterate the percentage of indoor water to be supplied with rainwater (Step 1.4 cell I33) and the percentage of irrigation to be supplied with rainwater (Step 1.5 cell B7). We find this a much quicker iteration approach than trying to fiddle with the fixtures themselves (for indoor) or the total required irrigation calculation (for outdoor). And remember: if you do this, you'll again have to back-solve which combination of fixtures and usage patterns meet the indoor rainwater volume you've set as your demand.

Follow the above methodology and you'll find the absolute maximum demand volume that will result in a balanced system for your fixed storage capacity.

However, do you actually want to use this maximum demand as your optimized design? No, you do not, as it will result in a zero or near-zero tank balance at least once every year. For the same reasons stated in Step 2.1, you want to decide what minimum storage is reasonable for you. Then set your minimum month-end storage volume in your mind and go back to your demand volumes. Slowly decrease the demands (Step 1.3–1.6) until the smallest number in all storage performance columns exceeds that minimum value.

You have now found the demands that, for the roof size, storage capacity, and average rainfall that you have defined, result in optimized storage performance.

Once again, you'll next want to ask yourself: *Is this design acceptable?*

Example: Optimize storage for Vancouver home (Part 2)

The Vancouver homeowners build the Part 2 table in their spreadsheet by following the Part 2 template. They are anticipating starting up their system in May, so they put the commission month formula, for Year 0, into the row associated with May.

Once the template is built, they pull out a copy of Figure 3.3 and they follow the step-by-step process.

Step 2.1 Back-Solve Storage Capacity

i. They first estimate a starting storage capacity volume by adding up the three largest demand volume months of June, July, and August and come up with approximately 64,000 liters (16,909 gal).

ii. They input 64,000 liters (16,909 gal) into cell D3 and look at the resulting storage performance (Shown as Step 2.1 - A)

iii. There are no months in Years 1–4 (columns G–J) with a zero volume, so they skip to (iv).

iv. They lower the storage capacity in small increments. At 50,000 liters (13,210 gal), they get zero month-end volumes in September. They then increase the storage capacity to 50,275 liters (13,283 gal), so that the smallest number in the storage performance columns is 25 liters (52 gal).[T]

v. At this point, they are not overly concerned about setting a minimum water volume in the tank.

Step 2.1 - A

	A ▶ ◀ E	F	G	H	I	J
1						
2						
3						
4						
5						
6						
7				Storage Performance		
8	Month	Net Supply	Year 0	Year 1	Year 2	Year 3	Year 4
9		liters [gal]	liters [gal]	liters [gal]	liters [gal]	liters [gal]	liters [gal]
10	Jan	19,984 [5,306]	0 [0]	64,000 [16,909]	64,000 [16,909]	64,000 [16,909]	64,000 [16,909]
11	Feb	15,177 [4,032]	0 [0]	64,000 [16,909]	64,000 [16,909]	64,000 [16,909]	64,000 [16,909]
12	Mar	14,797 [3,905]	0 [0]	64,000 [16,909]	64,000 [16,909]	64,000 [16,909]	64,000 [16,909]
13	Apr	9,420 [2,503]	0 [0]	64,000 [16,909]	64,000 [16,909]	64,000 [16,909]	64,000 [16,909]
14	May	6,456 [1,738]	6,456 [1,738]	64,000 [16,909]	64,000 [16,909]	64,000 [16,909]	64,000 [16,909]
15	Jun	-11,033 [-2,862]	0 [0]	52,967 [14,047]	52,967 [14,047]	52,967 [14,047]	52,967 [14,047]
16	Jul	-13,921 [-3,626]	0 [0]	39,046 [10,421]	39,046 [10,421]	39,046 [10,421]	39,046 [10,421]
17	Aug	-14,016 [-3,754]	0 [0]	25,030 [6,667]	25,030 [6,667]	25,030 [6,667]	25,030 [6,667]
18	Sep	-11,280 [-2,989]	0 [0]	13,750 [3,678]	13,750 [3,678]	13,750 [3,678]	13,750 [3,678]
19	Oct	14,930 [3,905]	14,930 [3,905]	28,680 [7,583]	28,680 [7,583]	28,680 [7,583]	28,680 [7,583]
20	Nov	27,470 [7,218]	42,400 [11,122]	56,150 [14,801]	56,150 [14,801]	56,150 [14,801]	56,150 [14,801]
21	Dec	24,069 [6,326]	64,000 [16,909]	64,000 [16,909]	64,000 [16,909]	64,000 [16,909]	64,000 [16,909]

Step 2.1 - A: *Storage capacity set to 64,000 liters for the first iteration. Note that columns B–D are not shown.*

Their Part 2 table is shown in Step 2.1 - B :

Step 2.1 - B

	A ▶◀ E	F	G	H	I	J
1						
2						
3						
4						
5						
6						
7			Storage Performance			
8 Month	Net Supply	Year 0	Year 1	Year 2	Year 3	Year 4
9	liters [gal]	liters [gal]	liters [gal]	liters [gal]	liters [gal]	liters [gal]
10 Jan	19,984 [5,306]	0 [0]	50,275 [13,283]	50,275 [13,283]	50,275 [13,283]	50,275 [13,283]
11 Feb	15,177 [4,032]	0 [0]	50,275 [13,283]	50,275 [13,283]	50,275 [13,283]	50,275 [13,283]
12 Mar	14,797 [3,905]	0 [0]	50,275 [13,283]	50,275 [13,283]	50,275 [13,283]	50,275 [13,283]
13 Apr	9,420 [2,503]	0 [0]	50,275 [13,283]	50,275 [13,283]	50,275 [13,283]	50,275 [13,283]
14 May	6,456 [1,738]	6,456 [1,738]	50,275 [13,283]	50,275 [13,283]	50,275 [13,283]	50,275 [13,283]
15 Jun	-11,033 [-2,862]	0 [0]	39,242 [10,421]	39,242 [10,421]	39,242 [10,421]	39,242 [10,421]
16 Jul	-13,921 [-3,626]	0 [0]	25,321 [6,795]	25,321 [6,795]	25,321 [6,795]	25,321 [6,795]
17 Aug	-14,016 [-3,754]	0 [0]	11,305 [3,041]	11,305 [3,041]	11,305 [3,041]	11,305 [3,041]
18 Sep	-11,280 [-2,989]	0 [0]	25 [52]ᵗ	25 [52]ᵗ	25 [52]ᵗ	25 [52]ᵗ
19 Oct	14,930 [3,905]	14,930 [3,905]	14,955 [3,957]	14,955 [3,957]	14,955 [3,957]	14,955 [3,957]
20 Nov	27,470 [7,218]	42,400 [11,122]	42,425 [11,175]	42,425 [11,175]	42,425 [11,175]	42,425 [11,175]
21 Dec	24,069 [6,326]	50,275 [13,283]	50,275 [13,283]	50,275 [13,283]	50,275 [13,283]	50,275 [13,283]

Step 2.1 - B: *A storage capacity of 50,275 liters [13,282 gal] has been back-solved as the smallest storage capacity that balances supply and demand.*

However, they don't put too much more thought into it as they already know that a 50,000 liter (13,210 gal) tank is far beyond the budget and space restrictions that they have. This design is not acceptable.

Increase Catchment?

They move on and consider the possibility of increasing the size of their catchment. They are planning on building a garage, and realize that they could include the garage roof as part of their catchment. The garage footprint is 55 m² (592 ft²). They go into the spreadsheet table for Step 1.2 and increase the total catchment area to 255 m² (2,745 ft²).

Step 2.1 Back-Solve Storage Capacity (again)

Returning to the Part 2 table, they back-solve once again for storage capacity (cell D3) and this time arrive at an optimized

storage capacity of 40,500 liters (10,700 gal). By increasing their roof by 25%, they have decreased the optimized tank size by 20%. However, this is still too big and too expensive.

As they can no longer increase catchment area, they must now choose an optimization approach. Looking closely at the storage performance, they notice that supply and demand are most mismatched in June, July, August, and September, which corresponds to the months they were intending on irrigating their garden. They decide to reduce and re-evaluate their outdoor rainwater demands before returning again to Step 2.1.

Reduce Demand

After evaluating their irrigation needs, and considering many of the options and practices in Chapter 1 for landscape water security, the family decides to lower their outdoor rainwater demand to 50% of what was calculated by the original Step 1.5 methodology.

Therefore, in the Step 1.5 table, cell B7, they change the percentage of irrigation to be supplied with rainwater to 50%. They then return to the Part 2 table and, once again, follow the instructions to back-solve for storage capacity in Step 2.1.

The minimum storage capacity that results in a balanced system is now substantially lower than before, at 10,500 liters (2,774 gal). The results of their third iteration is shown in Step 2.1 - C.

They are finally at a tank size that is within their budget expectations. They also note that if they increase the tank size to 15,140 liters (4,000 gal) (to match typical commercial tank size combinations) they end up with a minimum stored volume of 4,660 liters [1,288 gal] in September (up from 20 liters [62 gal]$^\tau$ in the following table), which is roughly 70% of their indoor water demand. This seems very reasonable and they decide that they are ready to move on to Part 3 of the feasibility stage.

For a video tutorial version of this example, head to www.essentialrwh.com)

Step 2.1 - C

	A ▶◀ E	F	G	H	I	J
1						
2						
3						
4						
5						
6						
7			Storage Performance			
8	Month	Net Supply	Year 0	Year 1	Year 2	Year 3	Year 4
9		liters [gal]	liters [gal]	liters [gal]	liters [gal]	liters [gal]	liters [gal]
10	Jan	27,252 [7,233]	0 [0]	10,500 [2,774]	10,500 [2,774]	10,500 [2,774]	10,500 [2,774]
11	Feb	21,123 [5,609]	0 [0]	10,500 [2,774]	10,500 [2,774]	10,500 [2,774]	10,500 [2,774]
12	Mar	20,639 [5,446]	0 [0]	10,500 [2,774]	10,500 [2,774]	10,500 [2,774]	10,500 [2,774]
13	Apr	13,783 [3,659]	0 [0]	10,500 [2,774]	10,500 [2,774]	10,500 [2,774]	10,500 [2,774]
14	May	10,004 [2,684]	10,004 [2,684]	10,500 [2,774]	10,500 [2,774]	10,500 [2,774]	10,500 [2,774]
15	Jun	-670 [-109]	9,334 [2,575]	9,830 [2,665]	9,830 [2,665]	9,830 [2,665]	9,830 [2,665]
16	Jul	-4,352 [-1,084]	4,982 [1,491]	5,478 [1,581]	5,478 [1,581]	5,478 [1,581]	5,478 [1,581]
17	Aug	-4,473 [-1,247]	509 [245]	1,005 [334]	1,005 [334]	1,005 [334]	1,005 [334]
18	Sep	-985 [-272]	0 [0]	20 [62]$^\tau$	20 [62]$^\tau$	20 [62]$^\tau$	20 [62]$^\tau$
19	Oct	20,808 [5,446]	10,500 [2,774]	10,500 [2,774]	10,500 [2,774]	10,500 [2,774]	10,500 [2,774]
20	Nov	36,797 [9,670]	10,500 [2,774]	10,500 [2,774]	10,500 [2,774]	10,500 [2,774]	10,500 [2,774]
21	Dec	32,460 [8,533]	10,500 [2,774]	10,500 [2,774]	10,500 [2,774]	10,500 [2,774]	10,500 [2,774]

τ - *The discrepancy between units becomes particularly large at the very small volumes. However, results are still well within the assumed accuracy of the calculation methodology. Note that 52 gallons / 13,283 gallons (storage capacity) = 0.4% and 62 gallons / 4,000 gallons (storage capacity) ≅ 2%.*

Calculations in each choice of unit were performed independently and values in earlier input fields rounded to a significant number of digits. The end result is a rounding discrepancy if you compare SI vs US Customary results.

Step 2.1 - C: *In the third iteration, they discover that reducing irrigation demands by 50% results in a significantly lower optimized storage capacity of 10,500 liters [2,774 gal].*

And the Iteration Continues (or Not) ...

Hopefully, you've now got a sense for how this works and how you can end up iterating, and iterating until you discover your optimized system.

There may be situations where you find yourself in an infinite loop. If that's the case, you've basically discovered that, for your budget, space, and rainwater demands there is no possible or feasible solution. Unless you are willing to make a concession at some point, such as increasing your roof catchment, increasing your storage, or reducing your rainwater demands, you won't find a feasible configuration for your RWH system.

Initial Tank Charge and Commission Month

There's one last thing to throw into the mix: we've included the ability to input *an initial tank charge* to your RWH system in the Part 2 Template (cell D5). Note that this consideration is far more important in primary supply or off-grid rainwater scenarios.

The idea here is that you can start your system with a full tank of water, rather than start your system with an empty tank. The way to do this is filling your tank from a backup system (municipal water or groundwater well)

or hiring a water-hauling company to deliver water. Note that if you are purchasing water from a water-hauling company, you'll want to minimize your cost per liter by purchasing a full truck load, which is typically 10,000 liters (2,600 gal) for the smaller water-hauling trucks and 20,000 liters (5,200 gal) for the larger ones. Also, if water-hauling is part of your contingency plan, you'll absolutely want to be planning your storage capacity with these delivery volumes in mind.

Why might you want to initially charge your tank? It's a little bit like heading on a road trip in your car. Depending on the distance to the next gas station (i.e. rainfall), you might choose to fill up before you go! In addition, with a full tank, you can test for leaks and piping connections during your commissioning process.

Lastly, the month you choose to commission your system can also have an impact on performance, especially in Year 0. However, substantially more advanced spreadsheet functions are required to make this work seamlessly. If you really want to optimize your startup based on commission month, we recommend that you head to www.essentialrwh.com to download our Essential Rainwater Harvesting Tool, which includes a drop-down menu for commission month.

Part 3. Test and Redesign For Low Rainfall

You've just designed a system that is optimized for average rainfall, and you should feel comfortable that as long as you get average rainfall, your RWH system will meet your rainwater demands. Now, you'll consider how this system will perform when you don't get average rain. If you skipped Chapter 2, it's time to go back and read the section on Managing Your Risk and particularly the sections on Minimum Rain Limit, Minimum Operating Goal, and Adaptive Strategies.

There are two critical decisions you have to make:

(i) *Minimum Rain Limit:* What is the worst-case drought scenario you want your system to be able to handle?

(ii) *Minimum Operating Goal:* What kind of storage performance do you expect in this worst-case scenario?

Think of it this way: When you design a system for average rainfall (Part 2), the goal is to ensure that your storage performance (columns G, H, I, J) is optimized. However, when you are evaluating a system at its minimum rain limit (Part 3), the goal is ensure that your storage performance (columns P, Q, R) meets your minimum operating goal.

There are two possible outcomes of Part 3 of the feasibility stage:

• At your minimum rain limit, your storage performance meets your minimum operating goal and no changes are needed.

• At your minimum rain limit, your storage performance does not meet your minimum operating goal. You must make changes to your design.

Compared to the methodology for Part 1 and Part 2, Part 3 is far more context specific and requires increased subjective judgement.

The Part 3 Template

The first thing to do is to build your Part 3 spreadsheet table.

Go back to your Part 2 table. Right after the last column in the table (column J), you'll want to add eight new columns (K-R) as well as the formulas shown in the Part 3 Template (see page 60.)

You may notice that this template, and the formulas contained within, are quite similar to the Part 2 Template. That's because in Part 3 we are repeating Part 2, but simply changing some of the input conditions.

Note that the decrease in rainfall is stated as a percentage in reduction from average in cell O3.

In column K, you calculate the resulting supply, based on the input in cell O3. We call this the *drought supply.*

Column L and M are titled "Adaptive Strategies." Remember that adaptive strategies are real-time behaviors or existing/planned backup supply that you employ to reduce your rainwater demand in the event of a shortage of rain. For now, don't worry too much about these columns as we'll go into more detail shortly.

The drought net supply is calculated in column O, and this is simply the drought supply minus the drought demand.

Lastly, the template has three columns for three years of storage (columns P, Q, and R). The formulas are nearly identical to those presented in the Part 2 Template for the month-end volumes. These three columns represent your drought storage performance and are tied to your minimum operating goal.

You may have noticed that when it comes to assessing how the system will perform in the event of drought, this template makes two fairly major simplifications:

• It assumes that the drought will occur as a percentage reduction of rainfall equally distributed over the course of the year.

Part 3 Template: Test and Redesign For Low Rainfall

	K	L	M	N	O	P	Q	R
1	Supply, Demand, and Storage at Minimum Operating Conditions							
2								
3	Reduction in Rain From Average, %							
4								
5								
6								
7						Drought Storage Performance		
8	Drought Supply	Adaptive Strategies		Drought Demand	Drought Net Supply	Drought Year 1[YY]	Drought Year 2	Drought Year 3[Ψ]
9	Volume	%	Volume	Volume	Volume	Month End Volume	Month End Volume	Month End Volume
10	=C10*(1-O3)		=D10*L10	=D10-M10	=K10-N10	=IF(J21+O10<0,0, IF(J21+O10>D3,D3, J21+O10))[Y]	=IP(P21+O10<0,0, IP(P21+O10>D3,D3, P21+O10))[ξ]	ξ
11	=C11*(1-O3)		=D11*L11	=D11-M11	=K11-N11	=IF(P10+O11<0,0, IF(P10+O11>D3,D3, P10+O11))[YY]	=IF(Q10+O11<0,0, IF(Q10+O11>D3,D3, Q10+O11))[YY]	YY
12	=C12*(1-O3)		=D12*L12	=D12-M12	=K12-N12	=IF(P11+O12<0,0, IF(P11+O12>D3,D3, P11+O12))	=IF(Q11+O12<0,0, IF(Q11+O12>D3,D3, Q11+O12))	
13	=C13*(1-O3)		=D13*L13	=D13-M13	=K13-N13	=IF(P12+O13<0,0, IF(P12+O13>D3,D3, P12+O13))	=IF(Q12+O13<0,0, IF(Q12+O13>D3,D3, Q12+O13))	
14	=C14*(1-O3)		=D14*L14	=D14-M14	=K14-N14	=IF(P13+O14<0,0, IF(P13+O14>D3,D3, P13+O14))	=IF(Q13+O14<0,0, IF(Q13+O14>D3,D3, Q13+O14))	
15	=C15*(1-O3)		=D15*L15	=D15-M15	=K15-N15	=IF(P14+O15<0,0, IF(P14+O15>D3,D3, P14+O15))	=IF(Q14+O15<0,0, IF(Q14+O15>D3,D3, Q14+O15))	
16	=C16*(1-O3)		=D16*L16	=D16-M16	=K16-N16	=IF(P15+O16<0,0, IF(P15+O16>D3,D3, P15+O16))	=IF(Q15+O16<0,0, IF(Q15+O16>D3,D3, Q15+O16))	
17	=C17*(1-O3)		=D17*L17	=D17-M17	=K17-N17	=IF(P16+O17<0,0, IF(P16+O17>D3,D3, P16+O17))	=IF(Q16+O17<0,0, IF(Q16+O17>D3,D3, Q16+O17))	
18	=C18*(1-O3)		=D18*L18	=D18-M18	=K18-N18	=IF(P17+O18<0,0, IF(P17+O18>D3,D3, P17+O18))	=IF(Q17+O18<0,0, IF(Q17+O18>D3,D3, Q17+O18))	
19	=C19*(1-O3)		=D19*L19	=D19-M19	=K19-N19	=IF(P18+O19<0,0, IF(P18+O19>D3,D3, P18+O19))	=IF(Q18+O19<0,0, IF(Q18+O19>D3,D3, Q18+O19))	
20	=C20*(1-O3)		=D20*L20	=D20-M20	=K20-N20	=IF(P19+O20<0,0, IF(P19+O20>D3,D3, P19+O20))	=IF(Q19+O20<0,0, IF(Q19+O20>D3,D3, Q19+O20))	
21	=C21*(1-O3)		=D21*L21	=D21-M21	=K21-N21	=IF(P20+O21<0,0, IF(P20+O21>D3,D3, P20+O21))	=IF(Q20+O21<0,0, IF(Q20+O21>D3,D3, Q20+O21))	

Y - Start in January of drought year 1 (column P). Add the December year 4 month end volume (J21) to the January drought net supply (O10).
YY - For February and onwards until the end of the year, add the prior month end volume to the current month drought net supply.

ξ - At the start of the 2nd & 3rd drought years (column Q & R), add the prior December month end volume (row 21) to the January drought net supply (O10).
Ψ - Continue on with the same formula patterns in drought year 3 (column R). The equations are not shown.

• It assumes that the drought will begin in January.

We acknowledge that these are gross over-simplifications of how drought could and will occur. However, trying to be precise about what a drought will look like might also be folly. The best thing we can do is to first acknowledge that we are trying to predict the unpredictable (rain, that is). Then, using broad strokes, we look at performance and we look for fragility in our design. Based on our risk tolerance, we can then add overall robustness and resilience to our design using the options that are possible and/or pragmatic.

And, of course, if you feel that you have a very good idea and confidence in the drought scenario that you want to future-proof yourself against, you could easily modify the spreadsheet as needed.

Before we dive into the nuts and bolts of the Part 3 process, let's go over some considerations for minimum operating goals and adaptive strategies for each of the typical scenarios: supplemental supply, primary supply, and off-grid supply.

Your Minimum Operating Goal

Establishing your minimum rain limit and your minimum operating goal is very context-specific and closely related to the amount of risk you are willing to take and the critical (or non-critical) nature of the water you are intending to harvest.

Also, when first starting on Part 3, you might be completely unsure as to what is a reasonable and pragmatic minimum operating goal for your budget. If that's the case, that's okay. As part of the Part 3 process you'll be doing some analysis that will help to establish a decent starting point.

But let's start by looking at three examples. Each one has different considerations that subsequently inform the example minimum operating goal statements.

Supplemental Supply

The designer of a mid-sized garden irrigation system has the following considerations:

• She grows a fairly substantial portion of food in her garden, and relies on this to offset her food costs.
• She much prefers to irrigate with rainwater, but can switch to the municipal water system if need be.
• Municipal water costs about \$3 per m^3 (1 cent per gallon).
• She expects her system to be able to supply rainwater throughout her irrigation months most years; but, because her municipal water is so cheap, she is not overly concerned if there is a particularly dry year and she runs out of rainwater.

Minimum operating goal: As long as rainfall is within 25%–30% of average, she wants her system to maintain balance.

Primary Supply

The designer of a primary rainwater-supply system has the following considerations:

• The system is designed to meet all indoor water loads during average rain conditions. The drinking and bathing loads are about 50% of the total rainwater demands.
• There is a backup groundwater well on the property, but the water has a very high mineral and salt content, so it is not high-quality water.
• The climate is arid and much of the rain comes in the early and late spring.

Looking at past drought and rainfall data, the designer decides that in the last 25 years, the worst drought brought a 45% reduction in annual rainfall. Two possibilities for minimum operating goals would be stated as follows:

#1: In the event of a 45% reduction in annual rainfall, the system must continue to meet all

drinking and bathing loads (i.e. 50% of the normal operating demands).

#2: In the event of a 60% reduction in annual rainfall, the system must meet all demands for at least six months.

Both minimum operating goal options could be appropriate, as they both properly state the expected storage performance at the minimum rain limit. The designer might choose to test both before deciding which one to design his/her system to.

It would also be wise for the designer to estimate the cost and alternative options for supplying backup water.

Off-grid Supply

The designer of an off-grid homestead has the following considerations:

- The water usage in the home has thoroughly been scrutinized from the onset. There is a composting toilet system, and extensive water conservation is built into the normal operating conditions of the system.

- In the event of extreme drought, the only alternative is to bring in trucked water, at a cost of $500+ per load.

Here, two possibilities for minimum operating goals could be stated like this:

#1: The system must continue to be balanced, even if there is a 75% reduction in rainfall.

#2: The system must supply one full year of water if there is a 75% reduction in rainfall.

Again, both versions of the minimum operating goal are stated properly. In the first version, the designer is requiring that there are no zero month-end volumes in all three drought years (columns Q, R, S). In the second version, the designer is requiring that there are no zero month-end volumes in the first drought year only (column Q). Note that the first minimum

operating goal likely requires a larger catchment area and storage capacity, and hence would have a larger upfront cost.

Using Adaptive Strategies

Adaptive strategies are real-time behaviors or supplementation that you intend on putting in place to meet your minimum operating goal. In the Part 3 template, you state the adaptive strategies that you intend on employing as a volume percentage of your normal operating demands. Note also that you state adaptive strategies as a total volume reduction in demand due to (1) supplementing water *and/or* (2) conserving water.

For instance, if you state a 50% adaptive strategy for a given month, you are saying that you intend on:

Supplementing 50% of your normal demands with another water source

OR

Conserving 50% of your normal demands through behavior

OR

Using a combination of the two approaches to ultimately reduce your rainwater demands by 50%.

When using adaptive strategies to meet your minimum operating goal, consider the following:

- You may not want to rely on adaptive strategies for lower reductions in rainfall (10%–20%) because drought is often only perceived in hindsight over long periods of time.

- On the other hand, if you intend on using and monitoring short- to long-term weather forecasts as part of your management and operation plan and subsequently adjusting water usage based on storage volumes, then adaptive strategies are very appropriate.

- If you have no way to import water and you have already minimized your demand for your average operating condition as much as possible, it may not be reasonable to expect a further behavior-based reduction in rainwater demand.

- If you know that you have a secondary water supply, you can use the adaptive strategies percentage (column L) and the resulting volume (column M) to calculate what it might cost you to make up the supply shortage with your alternate water system.

We'll reiterate one more time that the idea here is that if you are able to rely on adaptive strategies in a drought, you may convince yourself that you don't have to increase the size of your roof and/or increase the size of your tank, ultimately meaning a lower upfront capital cost for your system.

On the other hand, relying on this kind of behavior adaptability or water supplementation might not at all be acceptable for your system and your context. Your job as a designer is to be sure to use adaptive strategies appropriately.

The Part 3 Process

The step-by-step process for the third and last part of the feasibility stage is shown in Figure 3.4.

Each step is described in detail in the following sections.

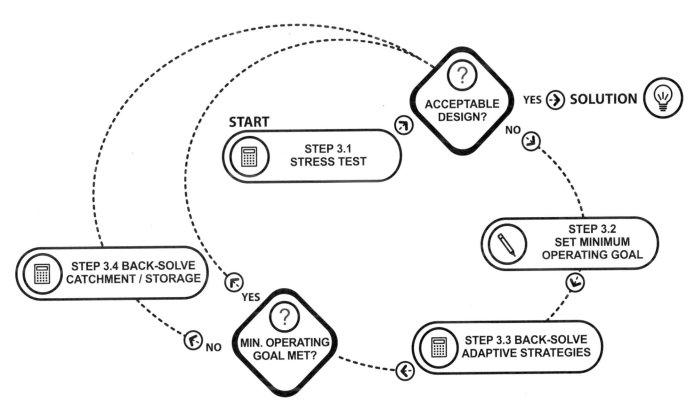

Fig. 3.4: *The step-by-step process for Part 3 of the feasibility stage: Test and Redesign for Low Rainfall.*
Credit: Verge Permaculture/ S. Fidler

Step 3.1: Stress Test

Stress testing is particularly useful tool, allowing you to test your system performance beyond its normal operating conditions to discover its limits. When stress testing, you'll start to form a clearer picture of how fragile or resilient your system is to reduced levels of rain. The only change you are making in the Stress Test step is to reduce the rainfall percentage in cell O3.

We recommend that you investigate two broad scenarios: a small-to-moderate drought over a longer time frame (say, three years) and a moderate-to-substantial drought over a shorter time frame (say one year, or a timeframe that is relevant to your climate).

Simply start by inputting a small reduction in rainfall into cell O3, such as 10%. See what happens. If your system immediately becomes unbalanced, your design is very fragile. Increase the reduction to 20%, then to 30%. Consider simultaneously what is happening to your water storage and how reasonable this reduction in rain might be for your climate. Increase the reduction to 40%, then to 50%. How quickly do you run out of water?

Once you get a general sense of your system performance and its limits, you'll want to move on to Step 3.2.

Acceptable Design?

The results of the Stress Test might actually give you confidence that your existing design is indeed perfectly acceptable. If so, you pass the Stress Test. Congratulations! You have found a solution to the feasibility stage and are ready to go onward with detailed design.

Alternatively, you may discover that under reduced rainfall conditions your drought storage performance is not acceptable. In that case, you'll move on to Step 3.2.

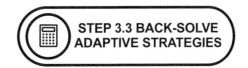

Step 3.2: Set Your Minimum Operating Goal

Here, you want set your minimum rain limit and your minimum operating goal.

There are some recommendations in Chapter 2 as well as at the beginning of this section for considering and establishing what these should be, but you'll also want to take into consideration the storage performance from your Stress Test. For instance, if you are already budget constrained and you notice that your system is unbalanced at a 20% reduction in rain, trying to insure yourself against a 50% reduction in rain is likely going to be cost prohibitive and unreasonable for you.

On the other hand, you might notice that your system becomes unbalanced at a 30% reduction in rain and wonder if perhaps meeting a 40% minimum rain limit might not be too far out of reach.

Remember to state your minimum operating goal such that you can translate it to storage performance, i.e. what you can see happening in your drought storage performance columns (P, Q, R) for your minimum rain limit.

Step 3.3: Back-Solve Adaptive Strategies To Meet Your Minimum Operating Goal

Before you go and add more capacity (and cost!) to your system, either by increasing your storage volume capacity or increasing the size of your catchment, we'll recommend that you first consider if it is appropriate to use adaptive strategies to effectively reduce your rainwater demands in order to meet your minimum operating goal.

By now you've defined the following:

- Minimum rain limit (cell O3)
- Minimum operating goal (the storage performance you expect in columns P, Q, and R)
- Storage capacity (cell D3)

Rather than iterate blindly, you'll want to consider the drought storage performance. Which months have zero month-end volumes? Do these months correlate with non-critical rainwater usage (such as irrigation)?

To then back-solve for adaptive strategies, input small percentage value(s) into cells L10:L21. Watch what happens to your drought storage performance. Increase the values in cells L10:L21 until you meet your minimum operating goal, without going over 100%.

Remember that the percentages that you input into column L can actually represent several different things. Use column L appropriately, depending on your context and scenario.

Adaptive Strategy: Behavior-Based Reduction in Demand

Use the percentages in column L and the subsequent volumes in column M to represent the reasonable behavior-based reduction in demand that you would like to account for when meeting your minimum operating goal. Are you willing to cut out water uses? Are you only trying to meet certain demands at your minimum rain limit?

Adaptive Strategy: Supplementation from Another Water Source

You can also use the percentages in column L to represent the volume of water that you will supplement with another water-supply system. Once you've back-solved the percentages in column L that meet your minimum operating goal, you can use the volumes calculated in column M to calculate your cost to supplement. You can use this information to help you decide if you want to rely on supplementation as a future operating cost, or if you'd prefer to put in a larger catchment area or a larger storage capacity as an upfront capital cost.

Design Strategy: Reduce Total Rainwater Demand

There's one last nifty way you can use the adaptive strategies column. The percentage in column L will also represent the amount that could reduce your Total Rainwater Demand (Step 1.3–1.6) in order to balance your system. Rather than decide to employ adaptive strategies, you could go back to Part 1 and start cutting out rainwater uses. The difference here is that now we are changing the design for the normal rainfall operating condition.

Minimum Operating Goal Met?

Once you've met the minimum operating goal by back-solving adaptive strategies, go on to *Acceptable Design?* described on the next page.

If you are not using adaptive strategies to meet your minimum operating goal, or if you are unable to back-solve adaptive strategies that meet your minimum operating goal, go on to Step 3.4.

Step 3.4: Back-Solve Catchment or Storage

Now you are defining the following:

- Minimum rain limit (cell O3)
- Minimum operating goal (the storage performance you expect in columns P, Q, and R)
- Adaptive strategies (cells L10:L21)

Note that the adaptive strategies could be defined as 0%, for instance if you were in an off-grid supply scenario with no possibility of backup, or if you don't want to rely on behavior changes to meet your minimum operating goal.

If possible, it's often more effective to start with increasing catchment area. Otherwise, increase the storage capacity (cell D3) in small increments until your storage performance (columns P, Q, R) meets your minimum operating goal.

Acceptable Design?

You may be in your second, third, or fourth iteration loop and are coming back to this question, having just back-solved for adaptive strategies (Step 3.3) or having just back-solved for catchment/storage (Step 3.4). Either way, you are now asking yourself if the new combination of catchment, storage capacity, and adaptive strategies meets your needs for your budget. If yes, you have found a solution and are ready to go onward with detailed design. If no, continue on and repeat Step 3.2.

If you are continuing on, note that something has to change. Unless you are willing to make a concession (such as reducing your minimum operating goal expectations), you'll be stuck in an infinite loop with no possible solution.

Example: Redesign for low rainfall for Vancouver home (Part 3)

The Vancouver homeowners follow the Part 3 template, and add columns K through R to their Part 2 table.

Next, they pull out a copy of Figure 3.4 to help them follow the step-by-step process for Part 3.

Step 3.1 Stress Test

Starting with a storage capacity of 15,140 liters (4,000 gal), they start their Stress Test and note the following for the drought years:

- 10% rain reduction: System is still balanced.
- 20% rain reduction: Month-end volumes hit zero in August and September.
- 30% and 40% rain reduction: Same result as 20% reduction.
- 50% rain reduction: Month-end volumes hit zero in July, August, and September.

Acceptable Design?

They decide that this performance is not acceptable. They'd like increased assurance that they could count on their RWH system, especially in the higher probability conditions of 10%–30% rainfall reductions from average.

Step 3.2 Set Minimum Operating Goal

They state their minimum operating goal as follows: They don't want to have to rely on their backup system (the municipal water) unless rainfall is less than 30% of average.

Step 3.3 Back-Solve Adaptive Strategies

They skip Step 3.3 because their minimum operating goal is about meeting their demands at the low rainfall condition (and not about conserving or supplying backup). They move on to Step 3.4.

Step 3.4 Back-Solve Catchment/Storage

They are now looking at adjusting either catchment (Step 1.2, cell B1) or storage capacity (cell D3) in an attempt to back-solve for a balanced system during the drought years. They've already increased catchment as much as possible, so they choose to iterate the storage capacity. They discover that it would take a storage capacity of 24,100 liters (6,367 gal) to balance their system in a 30% reduction in rain.

Acceptable Design?

Overall, they are not really happy with that size and cost of storage capacity. They do some research and decide that the absolute maximum storage capacity that they are willing to purchase is 18,925 liters (5,000 gal). They don't yet want to change their minimum operating goal (Step 3.2), so they go to Step 3.3.

Step 3.3. Back-Solve Adaptive Strategies

In Step 3.3, they set the storage capacity (cell D3) to 18,925 liters (5,000 gal) and the reduction in rain from average (cell O3) to 30%. They still have zero month-end volumes in August and September, as shown in Step 3.3 - A.

As such, they start back-solving the percentages in column L in attempt to understand the demand volumes that will result in a balanced system. After a few iterations, they realize that 10% adaptive strategies in each June, July, August, and September results in a balanced system (shown in Step 3.3 - B)

They realize that in order to balance their system at a 30% reduction in rain, they could do the following:

Adaptively conserve 10% of their demands when they hit a low rainfall condition.

OR

Adaptively supply (via the municipal system) 10% of their demands when they hit a low rainfall condition.

OR

Change their design strategy, go back to Part 1 (Steps 1.3–1.6) and cut out 10% of their normal operating condition demand loads.

They opt to go back to Step 1.5 and further reduce their outdoor rainwater demand volumes from 50% to 40% (Step 1.5 cell B7).

Step 3.3 - A: *Storage capacity set to 18,925 liters [5,000 gal] and reduction in rain from average at 30%.*

Step 3.3 - B: *Storage capacity set to 18,925 liters [5,000 gal] and reduction in rain from average at 30% and adaptive strategies set to 10% for each June, July, August and September.*

Step 3.3 - A

	K	L	M	N	O	P	Q	R
1	Supply, Demand, and Storage at Minimum Operating Conditions							
2								
3	Reduction in Rain From Average, %			0.3				
4								
5								
6								
7						Drought Storage Performance		
8	Drought Supply	Adaptive Strategies		Drought Demand	Drought Net Supply	Drought Year 1	Drought Year 2	Drought Year 3
9	Volume	%	Volume	Volume	Volume	Month End Volume	Month End Volume	Month End Volume
10	23,588 [6,255]	0	0 [0]	6,445 [1,702]	17,143 [4,553]	18,925 [5,000]	18,925 [5,000]	18,925 [5,000]
11	19,298 [5,118]	0	0 [0]	6,445 [1,702]	12,853 [3,416]	18,925 [5,000]	18,925 [5,000]	18,925 [5,000]
12	18,959 [5,004]	0	0 [0]	6,445 [1,702]	12,514 [3,302]	18,925 [5,000]	18,925 [5,000]	18,925 [5,000]
13	14,160 [3,753]	0	0 [0]	6,445 [1,702]	7,715 [2,051]	18,925 [5,000]	18,925 [5,000]	18,925 [5,000]
14	11,514 [3,071]	0	0 [0]	6,445 [1,702]	5,069 [1,369]	18,925 [5,000]	18,925 [5,000]	18,925 [5,000]
15	9,293 [2,502]	0	0 [0]	13,945 [3,684]	-4,653 [-1,182]	14,273 [3,818]	14,273 [3,818]	14,273 [3,818]
16	6,715 [1,820]	0	0 [0]	13,945 [3,684]	-7,230 [-1,864]	7,043 [1,954]	7,043 [1,954]	7,043 [1,954]
17	6,630 [1,706]	0	0 [0]	13,945 [3,684]	-7,315 [-1,978]	0 [0]	0 [0]	0 [0]
18	9,072 [2,388]	0	0 [0]	13,945 [3,684]	-4,873 [-1,295]	0 [0]	0 [0]	0 [0]
19	19,077 [5,004]	0	0 [0]	6,445 [1,702]	1,2632 [3,302]	12,632 [3,302]	12,632 [3,302]	12,632 [3,302]
20	30,269 [7,961]	0	0 [0]	6,445 [1,702]	23,824 [6,259]	18,925 [5,000]	18,925 [5,000]	18,925 [5,000]
21	27,234 [7,165]	0	0 [0]	6,445 [1,702]	20,789 [5,463]	18,925 [5,000]	18,925 [5,000]	18,925 [5,000]

Step 3.3 - B

	K	L	M	N	O	P	Q	R
1	Supply, Demand, and Storage at Minimum Operating Conditions							
2								
3	Reduction in Rain From Average, %			0.3				
4								
5								
6								
7						Drought Storage Performance		
8	Drought Supply	Adaptive Strategies		Drought Demand	Drought Net Supply	Drought Year 1	Drought Year 2	Drought Year 3
9	Volume	%	Volume	Volume	Volume	Month End Volume	Month End Volume	Month End Volume
10	23,588 [6,255]	0	0 [0]	6,445 [1,702]	17,143 [4,553]	18,925 [5,000]	18,925 [5,000]	18,925 [5,000]
11	19,298 [5,118]	0	0 [0]	6,445 [1,702]	12,853 [3,416]	18,925 [5,000]	18,925 [5,000]	18,925 [5,000]
12	18,959 [5,004]	0	0 [0]	6,445 [1,702]	12,514 [3,302]	18,925 [5,000]	18,925 [5,000]	18,925 [5,000]
13	14,160 [3,753]	0	0 [0]	6,445 [1,702]	7,715 [2,051]	18,925 [5,000]	18,925 [5,000]	18,925 [5,000]
14	11,514 [3,071]	0	0 [0]	6,445 [1,702]	5,069 [1,369]	18,925 [5,000]	18,925 [5,000]	18,925 [5,000]
15	9,293 [2,502]	0.1	1,395 [368]	12,551 [3,315]	-3,258 [-813]	15,667 [4,187]	15,667 [4,187]	15,667 [4,187]
16	6,715 [1,820]	0.1	1,395 [368]	12,551 [3,315]	-5,835 [-1,496]	9,832 [2,691]	9,832 [2,691]	9,832 [2,691]
17	6,630 [1,706]	0.1	1,395 [368]	12,551 [3,315]	-5,920 [-1,609]	3,912 [1,082]	3,912 [1,082]	3,912 [1,082]
18	9,072 [2,388]	0.1	1,395 [368]	12,551 [3,315]	-3,479 [-927]	433 [155]	433 [155]	433 [155]
19	19,077 [5,004]	0	0 [0]	6,445 [1,702]	12,632 [3,302]	13,065 [3,457]	13,065 [3,457]	13,065 [3,457]
20	30,269 [7,961]	0	0 [0]	6,445 [1,702]	23,824 [6,259]	18,925 [5,000]	18,925 [5,000]	18,925 [5,000]
21	27,234 [7,165]	0	0 [0]	6,445 [1,702]	20,789 [5,463]	18,925 [5,000]	18,925 [5,000]	18,925 [5,000]

Returning again to Step 3.3, they see that their system is now balanced during all drought years with 0% adaptive strategies employed. This is shown in Step 3.3 - C.

At this point, they are quite happy with the performance of their system at their lower rainfall operating condition of 30%. As one last check, they decide to estimate the make-up volumes that they will have to supply in the event of a 40% reduction in rain. They increase cell O3 to 40% and see that only September has a zero month-end volume. They back-solve the adaptive strategy percentage for September (cell L18)

Step 3.3 - C

	K	L	M	N	O	P	Q	R
1			Supply, Demand, and Storage at Minimum Operating Conditions					
2								
3		Reduction in Rain From Average, %			0.3			
4								
5								
6								
7						Drought Storage Performance		
8	Drought Supply	Adaptive Strategies		Drought Demand	Drought Net Supply	Drought Year 1	Drought Year 2	Drought Year 3
9	Volume	%	Volume	Volume	Volume	Month End Volume	Month End Volume	Month End Volume
10	23,588 [6,255]	0	0 [0]	6,445 [1,702]	17,143 [4,553]	18,925 [5,000]	18,925 [5,000]	18,925 [5,000]
11	19,298 [5,118]	0	0 [0]	6,445 [1,702]	12,853 [3,416]	18,925 [5,000]	18,925 [5,000]	18,925 [5,000]
12	18,959 [5,004]	0	0 [0]	6,445 [1,702]	12,514 [3,302]	18,925 [5,000]	18,925 [5,000]	18,925 [5,000]
13	14,160 [3,753]	0	0 [0]	6,445 [1,702]	7,715 [2,051]	18,925 [5,000]	18,925 [5,000]	18,925 [5,000]
14	11,514 [3,071]	0	0 [0]	6,445 [1,702]	5,069 [1,369]	18,925 [5,000]	18,925 [5,000]	18,925 [5,000]
15	9,293 [2,502]	0	0 [0]	12,445 [3,287]	-3,153 [-785]	15,773 [4,215]	15,773 [4,215]	15,773 [4,215]
16	6,715 [1,820]	0	0 [0]	12,445 [3,287]	-5,730 [-1,468]	10,043 [2,747]	10,043 [2,747]	10,043 [2,747]
17	6,630 [1,706]	0	0 [0]	12,445 [3,287]	-5,815 [-1,581]	4,228 [1,166]	4,228 [1,166]	4,228 [1,166]
18	9,072 [2,388]	0	0 [0]	12,445 [3,287]	-3,373 [-899]	855 [267]	855 [267]	855 [267]
19	19,077 [5,004]	0	0 [0]	6,445 [1,702]	12,632 [3,302]	13,487 [3,569]	13,487 [3,569]	13,487 [3,569]
20	30,269 [7,961]	0	0 [0]	6,445 [1,702]	23,824 [6,259]	18,925 [5,000]	18,925 [5,000]	18,925 [5,000]
21	27,234 [7,165]	0	0 [0]	6,445 [1,702]	20,789 [5,463]	18,925 [5,000]	18,925 [5,000]	18,925 [5,000]

Step 3.3 - C: *Storage capacity set to 18,925 liters (5,000 gal) and reduction in rain from average at 30% and outdoor rainwater demand reduced to 40% of total irrigation requirements. The system is balanced with a minimum volume of 855 liters (267 gal) in the month of September.*

until the month-end volume is no longer zero, as shown in Step 3.3 - D.

They now know that, in the event of a severe drought (defined, in their case as a 40% long-term reduction from average), they might have to supplement around 4,000 liters (1,000 gal) in the month of September (taken from cell M18). They can now estimate the cost of this supplementation using the water fees charged to them by their municipality.

For a video tutorial version of this example, head to www.essentialrwh.com.

Step 3.3 - D

	K	L	M	N	O	P	Q	R
1	Supply, Demand, and Storage at Minimum Operating Conditions							
2								
3	Reduction in Rain From Average, %			0.4				
4								
5								
6								
7						Drought Storage Performance		
8	Drought Supply	Adaptive Strategies		Drought Demand	Drought Net Supply	Drought Year 1	Drought Year 2	Drought Year 3
9	Volume	%	Volume	Volume	Volume	Month End Volume	Month End Volume	Month End Volume
10	20,218 [5,361]	0	0 [0]	6,445 [1,702]	13,773 [3,659]	18,925 [5,000]	18,925 [5,000]	18,925 [5,000]
11	16,541 [4,387]	0	0 [0]	6,445 [1,702]	10,096 [2,685]	18,925 [5,000]	18,925 [5,000]	18,925 [5,000]
12	16,250 [4,289]	0	0 [0]	6,445 [1,702]	9,805 [2,587]	18,925 [5,000]	18,925 [5,000]	18,925 [5,000]
13	12,137 [3,217]	0	0 [0]	6,445 [1,702]	5,692 [1,515]	18,925 [5,000]	18,925 [5,000]	18,925 [5,000]
14	9,869 [2,632]	0	0 [0]	6,445 [1,702]	3,424 [930]	18,925 [5,000]	18,925 [5,000]	18,925 [5,000]
15	7,965 [2,145]	0	0 [0]	12,445 [3,287]	-4,480 [-1,143]	14,445 [3,857]	14,445 [3,857]	14,445 [3,857]
16	5,756 [1,560]	0	0 [0]	12,445 [3,287]	-6,689 [-1,728]	7,756 [2,130]	7,756 [2,130]	7,756 [2,130]
17	5,683 [1,462]	0	0 [0]	12,445 [3,287]	-6,762 [-1,825]	994 [305]	994 [305]	994 [305]
18	7,776 [2,047]	0.3	3,734 [986]	8,712 [2,301]	-936 [-254]	59 [51]τ	59 [51]τ	59 [51]τ
19	16,352 [4,289]	0	0 [0]	6,445 [1,702]	9,907 [2,587]	9,965 [2,638]	9,965 [2,638]	9,965 [2,638]
20	25,945 [6,823]	0	0 [0]	6,445 [1,702]	19,500 [5,121]	18,925 [5,000]	18,925 [5,000]	18,925 [5,000]
21	23,343 [6,141]	0	0 [0]	6,445 [1,702]	16,898 [4,439]	18,925 [5,000]	18,925 [5,000]	18,925 [5,000]

τ — The discrepancy between units becomes particularly large at the very small volumes. However, results are still well within the assumed accuracy of the calculation methodology. Note that 51 gallons / 5,000 gallons (storage capacity) = 1%.

Calculations in each choice of unit were performed independently and values in earlier input fields rounded to a significant number of digits. The end result is a rounding discrepancy if you compare SI vs US Customary results.

Step 3.3 - D: *Storage capacity set to 18,925 liters (5000 gal) and reduction in rain from average at 40%. Cell L18 back-solved to 30% to ensure drought storage performance greater than zero.*

The Feasibility Stage Summarized

To summarize, you'll want to do the following to complete the feasibility stage for your RWH system.

Build your calculation tool by following all of the templates provided, or download ours.

Start with Part 1 step-by-step process shown in Figure 3.1 along with the descriptions for each step. Define your supply and demand at average rainfall conditions and iterate until you have a supply-to-demand ratio that is greater than 1.0.

Move on to Part 2 to optimize your storage for the catchment area and the demands you defined in Part 1. Follow the step-by-step process in Figure 3.3 along with the descriptions for each step. Iterate until you have an acceptable design for your needs and your budget.

Finish with Part 3 and evaluate the system that you designed in Part 2 in low rainfall conditions. Follow the step-by-step process in Figure 3.4 along with the descriptions for each step. Make sure that your system is robust enough to meet your minimum operating goal; if not, make the necessary changes. Iterate until you have an acceptable design for you needs and your budget.

The example provided in this chapter is also included on our website in a video tutorial format. In addition, there you'll find several other step-by-step examples of the three-part feasibility process, including the design of a fully off-grid system.

SOLUTION

Your Solution

Once you have completed the feasibility stage, you will have the following:

- Catchment area
- Storage capacity
- Rainwater demand volumes

You are now ready to move on with more comprehensive planning; this will include your site layout (Chapter 4), the detailed design of your collection (Chapter 5), storage (Chapter 6), and pump components (Chapter 7).

Chapter 4

Anatomy of a Rainwater System

As anatomy is the study of the structure and the parts of an organism, this chapter is about the essential components that make up a RWH system and their relative placement, as well as spatial layout and positioning considerations.

You can think of relative placement as the order in which a drop of water might move through each of the components. Your spatial layout is the location of the RWH system components relative to your house and your landscape. You'll also need to think about elevation because much of the successful design of a RWH system depends on the ability of water to passively flow from one component to the next.

This chapter is fundamentally about making sure that the individual components that make up your system work well together and the system functions as a whole. Remember, a properly designed and maintained RWH system will behave as a treatment train — this is the end goal.

The Essential Components

There are many individual components that make up a rainwater system.

Collection and pre-filtration is about capturing rain, diverting the rain, and removing the larger debris. In this category, the biggest difference between anatomies in RWH systems is often driven by the choice of pre-filtration strategies and the subsequent required components and placement of those components.

As for *storage*, all tanks will have pretty much the same components: inlets, outlets, overflows, air vents, etc. When considering *distribution*, you'll start to get far more context-specific, depending on the supply scenario. Are you supplying irrigation or indoor plumbing? Do you require pressurized water? Where do you require that water?

Figure 4.1 shows the typical components for a RWH system. We'll be referring to this figure quite often, so you may want to flag it so you can return to it quickly and easily.

Collection and Pre-Filtration Components:

1. **Roof:** Your roof is your rainwater collection surface (the *catchment* area).
2. **Rain gutters:** Rain gutters convey water to the downspouts.
3. **Downspouts:** The downspout is the vertical conveyance that takes the water away from the rain gutters.
4. **Diversion valve:** A diversion valve is optional — more frequently used in seasonal cold-climate systems.

5. **Diverted water and components:** This is the corresponding downspout, conveyance piping, and water that is diverted away from a tank by a diversion valve.
6. **Pre-filtration:** A pre-filtration device filters out larger debris such as leaves, bird droppings, and other vegetation. This illustration shows a downspout filter. Other options and pre-filtration strategies are discussed in Chapter 5.
7. **First flush device:** This optional device might be included as part of a pre-filtration

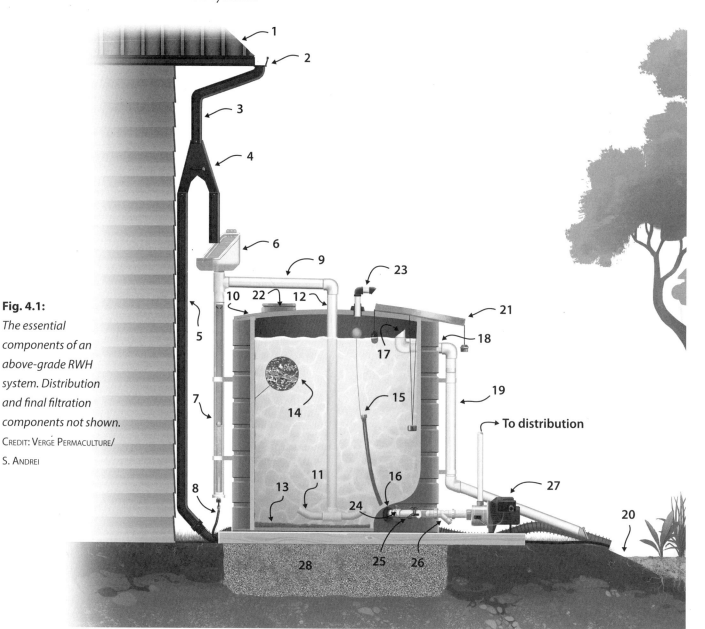

Fig. 4.1:
The essential components of an above-grade RWH system. Distribution and final filtration components not shown.
CREDIT: VERGE PERMACULTURE/
S. ANDREI

strategy. It is a vessel that diverts the first volume of rain in a rain event.

8. **First flush drainage:** The port through which the first flush chamber is drained.

9. **Conveyance piping:** Pipe and associated fittings that get the water to your rain tank. The configuration and amount of pipe you'll need is dependent on the placement of your tank relative to your gutters and downspouts.

Storage and Distribution Components:

10. **Tank:** Your water storage vessel, or vessels. The total volume that you can store in the tank(s) is your storage capacity.

11. **Quiet Inlet:** A well-designed inlet ensures that water entering the system does not disturb the sludge on the bottom of the tank. As shown, it can be designed to "point" the water up instead of down, and be split into two streams, de-energizing the water entering the tank.

12. **Inlet Port:** This fitting allows for the inlet connection to pass through the tank wall without allowing any leaks.

13. **Sludge layer:** This is a layer at the bottom of the tank that develops with time and consists of settled debris and anaerobic microbes.

14. **Biofilm:** This thin film layer is comprised of a community of microbes and forms on the internal walls of the tank and on the sludge layer interface.

15. **Intake:** A well-designed intake also minimizes disruption in the water column and ensures that the outlet water is as clean as possible. Your intake can either have a static/fixed outlet or a floating outlet (shown) and may be equipped with a screen.

16. **Outlet port:** This fitting allows for the outlet connection to pass through the tank wall without allowing any leaks.

17. **Overflow inlet:** The height of the pipe inlet will set the maximum water level in the tank.

18. **Overflow port:** This fitting allows for the overflow connection to pass through the tank wall without allowing any leaks.

19. **Overflow piping:** Water beyond the maximum capacity of the tank is sent either to landscape overflow (shown) or, alternatively, to another tank.

20. **Water-harvesting earthworks:** At a minimum, overflow water must be directed away from all foundations. Preferably, this water is directed to other water-harvesting features such as rain gardens, swales, etc.

21. **Level indicator:** This is a device that you can fit to your tank to indicate level.

22. **Manway:** A hatch that is used for access and maintenance, typically large enough for an adult to enter.

23. **Air vent:** As water enters or leaves, the tank needs to be able to replace that water volume with air. Most commercial rainwater tanks have integrated air vents in the manway to provide this function.

24. **Isolation valve:** A valve for isolating the tank.

25. **Hose bib:** An exterior faucet provides easy access to the tank water.

26. **Y strainer:** If a pump is used, install a pre-filter upstream to protect the pump. Alternatively, a screen can also be installed on the intake.

27. **Pump:** The pump can be external to the tank (shown) or be a submersible model located inside the tank.

28. **Foundation:** All tanks require a solid foundation.

Building a Site Plan

There are three very good reasons why you should build a site plan.

The first is that a good site plan is probably one of the most important tools to help you with the proper design and spatial layout of your system.

Secondly, a site plan serves as an indispensable resource during construction. Combined with construction drawings, you can effectively and very simply convey information to contractors or others helping you during installation.

Lastly, your site plan can serve as the basis for capturing your final construction and *as-built* notes. This kind of recordkeeping is very important for future maintenance, upgrades, and changes; it is particularly important if you install any underground infrastructure.

Size and Scale

Your site plan should be drawn to scale. That means that the proportional ratios on your paper are representative of the real-life proportions. If you are doing your drawings by hand (as we often do), an architect's scale ruler is an indispensable tool. It allows you to easily convert your real-life dimensions onto a piece of paper using common paired scales (1:100, 1:200, 1:300, and so on). If you are using an architect's scale with imperial units, these are typically marked with a ratio of x″= y′, where x is the measurement on the paper drawing in inches and y is the number of feet in real life. See Figure 2.2 for a picture of an architect's scale ruler.

When choosing a scale, pay attention to not fill the whole sheet of paper with the drawing. We find that ledger-sized graph paper (279 × 432 mm [11 × 17 in]) is usually the ideal size for both clearly communicating the system components at an appropriate scale and leaving enough room for notes and comments in the margins.

When the site plan is drawn to scale, horizontal distances can be measured directly off the drawing using a scale ruler, saving a ton of time when planning and designing your system. To see an example of this, skip ahead briefly to Example 5.1.

Getting Started

Your site plan describes your site layout and positioning of your components and anything else that is relevant to the design. It is drawn as a bird's eye view — as the site would appear if looking down at it from above.

You'll want to start with your property boundaries and property dimensions as well as a compass rose to indicate the cardinal direction.

You may have to simply go out and gather this information on your site using a tape measure and boundary markers, recording the dimensions and relative placement of all the infrastructure that exists and is relevant to your design. However, you'll be able to save a lot of time if you are able to build your site plan by starting with a base map.

Using a Base Map

Starting with a base map can be a huge time-saver, especially for sites with a lot of existing infrastructure. Rather than start your site plan from scratch, it's well worth first looking for resources that can be easily used as a base map for your site plan.

The most commonly available resources are the following:

Cadastral map. These are produced by your municipality and are used for tax purposes. They are produced by professional surveyors and show boundaries, easements, property improvements, roads, etc. Other names include: *real property report, survey certificate,* and *survey report.* See Figure 4.2 for an example of a base-map made from a municipal cadastral map.

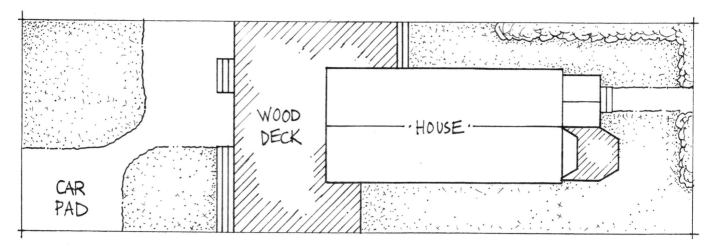

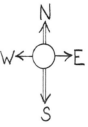

Aerial photography. If you have access to an aerial photo of your site, you can use this as a base map. However, before doing so, you'll need to determine the scale of the photo and convert or match it to the scale you intend on using for your site plan. The easiest way to do that is to take the actual measurement between two points on the photograph and the measured ground distance between those same two points.

Photograph scale = the photo distance ÷ the ground distance.

Architectural drawings. If you have architectural drawings of your home, you can use the plan view drawing produced by the architect.

There's always the option of hiring a professional surveyor to produce a base map (i.e. full-property map) for you. Although the cost of this would likely not be warranted for the design of a simple RWH system, it can be a very valuable (and necessary) resource if you are producing a full water-security plan for your property, especially for acreages and farms.

Using Mapping and CAD Software

There are a few popular software programs that can help you produce your base map, notably Google Earth Pro, which is free.

First off, Google Earth Pro can likely provide you with a high-definition aerial photo of your property, no matter where you are in the world (with the exception of some rural or low-density population locations). You can simply export a photo to any number of third-party programs and then determine its scale using the method described in the section above. Computer Aided Design (CAD) programs such as SketchUp or Smartdraw will then allow you to increase or decrease the scale to appropriately fit whatever sized paper you are using.

Consider how much time this process may save you, particularly if there is a lot of existing infrastructure on your site. Instead of having to measure out and redraw every building, access way, boundary line, etc., you simply use an aerial photo and determine its scale using one ground measurement.

Note also that Google Earth Pro will also allow you to draw simple shapes and create a simple site plan directly within the program. If your needs are simple, this could work quite well for you. The program can also measure lengths as well as calculate areas. However, you should still confirm the accuracy of the Google Earth scale and its measurements by comparing them with a measurement taken from your actual site.

We usually prefer to export the Google Earth Pro aerial photo to a third-party program because it provides more flexibility for setting up

Fig. 4.2:
An example of a simple base map made from a cadastral map.

the base map and ultimately how we communicate and present the site plan for the property.

Data Overlay

Once again, your site plan needs to include everything that is relevant to your design. The different resources available to you (cadastral maps, aerial photography, architectural drawings) will provide you with some preliminary information about your site; however, there will be important information that is missing.

Your next step is to start adding more information to your site plan. There are four main categories of information: *elevations, sectors, access,* and *structures.* You can imagine that your site plan might start to become rather crowded with information. For that reason, site plans are usually created in layers, with the different categories of information provided on different layers. This is called *data overlay.*

Creating data overlays is very easy to do using CAD software. If you are building your map by hand, using multiple layers of translucent paper (also called *vellum*) allows you to keep individual layers uncluttered, while also allowing you to see everything at once, if need be.

Following is a summary of the data overlays you'll want to include on your site plan.

Elevation Overlay

For larger sites, such as acreages and farms, you'll absolutely want to indicate elevations (via contour lines) as a data overlay. How to create and read contour maps is outside the scope of this book, but know that understanding the landform is crucial to understanding how water will gravity flow across your site.

As onsite water flow is fundamentally tied to elevation and contour, you can also think of this data layer as your landscape water flow layer. Where does stormwater go? How does it flow across your property? If you have existing downspouts, you'll also want to indicate where they are discharging water and where that water subsequently flows.

Also, ask yourself: Are there existing water erosion issues? Drainage ditches? Steep slopes? Does water flow toward or away from building foundations? Where are the high points on the property? Where are the low points? Where are the wet spots or places that water accumulates?

We can't emphasize enough that energy-efficient and effective design is water-centric. Don't skip this step — even if you have no intention of providing landscape irrigation with your RWH system. This data overlay is crucial when considering tank placement and the subsequent tank overflows. See Figure 4.3 for an example of an elevation overlay.

Sector Overlay

Sectors are energies that interact with or cross your site. Examples of sectors relevant to RWH system include wind, view, fire, solar (summer

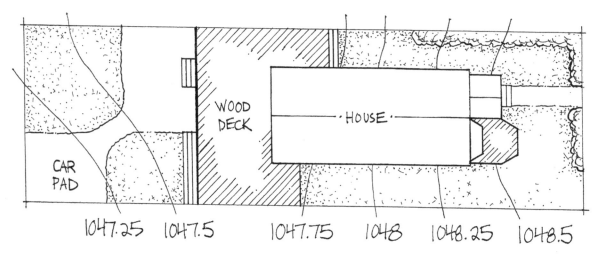

Fig. 4.3: *Contour lines (elevation overlay) shown on the base map. Elevation in meters.*

1047.25   1047.5   1047.75   1048   1048.25   1048.5

and winter sun), shade, wildlife, traffic, fire, and frost. See Figure 4.4 for an example of a sector overlay.

Access Overlay

An access overlay provides a way to represent and analyze the natural use of your property by the occupants and represents the well-worn natural flow lines (aka desire lines) around the yard and structures, in addition to roads and pathways required for site access.

The flow lines can be marked with arrows representing the traffic pathways and give you a visual reminder to avoid building or designing

elements that would interfere with or block them. See Figure 4.5 for an example of an access overlay.

Structures Overlay

Structures are elements in both the built and natural environment. Basically, on this data overlay you'll identify anything tangible and relevant to your design that is not already included on your base map or in any of the other layers.

In addition to including the building footprint(s), you'll want to make some notes:

• What is the existing roof material? What kind of shape is it in?

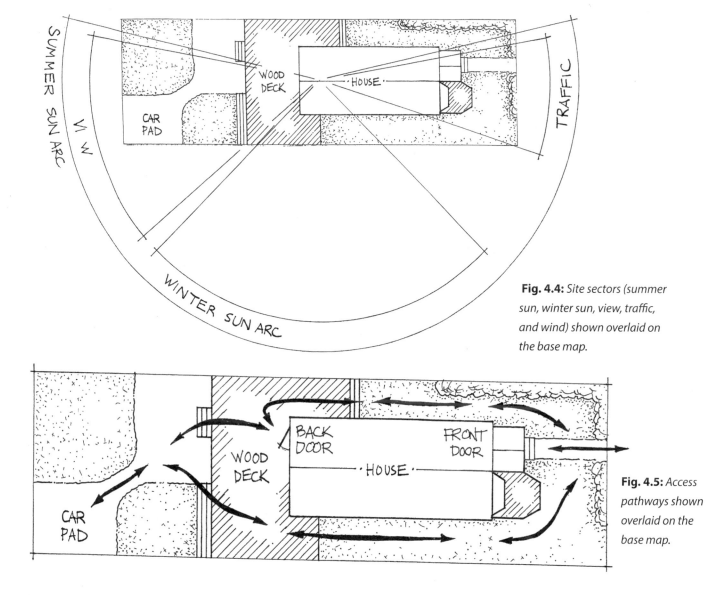

Fig. 4.4: *Site sectors (summer sun, winter sun, view, traffic, and wind) shown overlaid on the base map.*

Fig. 4.5: *Access pathways shown overlaid on the base map.*

- Are there trees or other overhanging vegetation over the roof?
- If gutters and downspouts are pre-existing, what materials are they made from? Where are the locations of the existing downspouts?
- What are the heights of your rain gutters relative to the ground level? What are the slopes of the existing gutters?

Make sure you indicate any existing water sources, such as wells and associated underground piping, and, if applicable, your septic field.

You'll absolutely want to know where any other underground utilities are located on your property, as this is crucial for an effective design and a safe installation. Seek out the organization in your jurisdiction that maintains a listing of buried facilities. In Canada and the US this service is called *one call,* and the website www.clickbeforeyoudig.com provides information for all Canadian and American provinces and states. Note, however, that these services do not usually list private infrastructure such as private water, power, or gas lines. If you have any suspicion that there may be unmarked or unregistered underground infrastructure, you may want to track down previous owners, talk to neighbors, or local contractors, or even rent an underground cable, pipe, and/or utility locator.

Depending on what you intend to use your captured rainwater for, you'll also want to indicate things like the actual (or proposed) irrigation location or any existing or planned plumbing that you may need to connect into. If you already know that you will be using a pump, you'll want to identify the places where you can access power.

You'll certainly include everything that is relevant and that already exists, but you may not want to stop there. This is a good opportunity to also consider any future infrastructure that you may want to place on your property.

Thinking About Spatial Layout

Having completed your initial site plan, next you'll typically evaluate the potential locations for your water storage. You want to start with your storage location because your tanks are the largest component(s) of the system and often have the most constraints in terms of where they can and cannot go.

The great thing about using data overlay as a design technique is that, as you start to add more layers of information to your site, the site naturally constrains itself; there are often only a couple of areas where it makes sense to place your storage. The better you get at this, the easier design gets and the less likely you are to make critical errors in judgement regarding placement of your system components.

Potential Storage Locations

You'll want to weigh your ideal storage location(s) with the following design considerations:

Elevation Considerations

- **Elevation pressure:** The higher the elevation you choose to place your tank relative to the elevation of your end-use, the more water pressure will be available to you; a large amount of elevation difference can minimize or even eliminate the need for a pump.

- **Elevation of your gutters relative to your tank:** You'll need to place your tank at an elevation such that water can gravity flow from your gutters into your tank. You'll also need to account for any vertical space needed for your pre-filtration strategy.

- **Foundation:** The tank foundation should shed water. Do not put your tank in a spot where water naturally pools or stagnates.

- **Water-harvesting earthworks:** The water that discharges from the overflow piping must be able to gravity flow away from the tank foundation. Ideally, this overflow is put to productive use through appropriate water-harvesting earthworks (swales, diversion drains, and rain gardens, etc.).

- **Failure:** In the event of tank failure, a wall rupture, or a major leak, consider where the water would go. Select a location (or excavate) such that this risk is minimized and the gravity flow of water is away from buildings, foundations, and infrastructure.

Sector Considerations

- **Shade:** Minimizing direct sunlight on the tank and keeping water temperature low will improve your water quality and extend the life of a plastic tank.
- **View:** Do you want to hide the tank from view?
- **Wind:** Plastic tanks in particular are susceptible to damage from wind. Your tank should be properly secured.

Access Considerations

- **Delivery:** If you plan on occasionally supplementing with trucked water, you'll need a location that is accessible to a delivery vehicle.
- **Servicing and inspection:** You'll need access for safe and periodic servicing and inspection.
- **Ease of monitoring:** Placing your tank along an already frequented route (such as a pathway) will improve the ease with which you can provide monitoring and maintenance.

Structure Considerations

- **Underground infrastructure:** Avoid placing a tank on top of any underground infrastructure. You don't want to eliminate the possibility of future servicing or repair to that infrastructure.
- **Set-backs:** Ensure proper spacing from the property lines, other buildings, other infrastructure, etc.

- **Ease of construction:** You'll want enough space not only for the tank footprint, but to construct the foundation. In really tight spaces, you will need to make sure there is even a way to get the tank into your proposed location.
- **Other foundation considerations:** Some locations, surfaces, and soils make for poor tank foundations.
- **Relative location to end-use:** Think about your end-use, including the location of the irrigation area, location of the plumbing tie-in(s), and power.
- **Existing downspouts:** If you are hoping to reuse existing gutters and downspouts, you'll also need to consider their location and height relative to potential tank locations.

Stacking Functions

There also may be an opportunity to plan your tank location such that it does more than simply hold water, i.e. *stacking functions*. Some additional questions that you might ask yourself:

- Could you use the tank to shield an unpleasant view?
- Can the tank provide wind protection to you, your garden, or any other property element that might benefit from such?
- Can the tank be designed such that it integrates with access or a pathway? (See Figure 4.6 for an example of this.)
- Can it be used as a trellis? Or provide support to other landscape elements?

We've seen tons of great examples of tanks serving multiple functions. The only limit here is your creativity.

Below-Grade Infrastructure

Below-grade installation usually adds considerable complexity to your design. When planning

Fig. 4.6: *This rain cistern (at Tagari Garden Farm in Tasmania) is stacking functions: it is acting as a rain storage and as a structural support for a catwalk access to the building.*

the layout of underground infrastructure, you have these additional considerations:

- Jurisdiction-specific codes and regulations pertaining to both technical and safety requirements.
- Site-specific soil considerations such as soil saturation, water table, and subsoil appropriateness.
- Site-specific access considerations for installation and construction.
- Manufacturer-specific requirements such as maximum buried depth, foundation, etc.

The large number of context-specific considerations and the possibility for technical complexity are far beyond the essentials of RWH system design. As such, if you are considering an underground tank, skip to Chapter 6 where we provide guidance to help you decide if you might explore this direction further with your local professionals.

Overflow Water and Water-Harvesting Earthworks

We've already emphasized the importance of ensuring that the water from your tank overflow is directed away from all foundations. However, rather than send that water directly off site or to the storm sewer, we recommend that you view the water that overflows from your tank as simply another resource waiting to be put to productive use. Through careful consideration of elevations and the creation of some moderate earthworks, water can be circulated, stored, and redirected to where it is useful. As such, any opportunities for landscape water harvesting should be included in your consideration of tank placement.

It is this kind of integrated design approach that will ultimately deliver increased productivity, water security, and resilience.

Single vs Multiple Tanks

After considering all of the above, you may only have one ideal storage location, but usually there are at least a few options. Plus there are a number of possible permutations when it comes to actually selecting and placing tanks. You might have:

- a single tank at a single location
- multiple tanks at a single location
- multiple tanks at multiple locations

In Figure 4.7, the gutters, their slope, the existing downspouts, and potential tank locations as well as their relative distances have been identified.

Whether you are using a single tank at a single location or multiple tanks at a single location is a question of tank capacity, tank sizes, installation or space limitations, and what is commercially available. For instance, if you are looking for a total storage capacity of 19,000 liters (5,000 gal), it might be easier to fit in two 8,000 liter (2,500 gal) tanks.

There are a few possible motivations when considering whether to use one vs two locations (or more). One major reason for needing to go with multiple locations might be that it is too difficult to move all of the water from your roof(s) to one central location. Basically, your rain gutters and conveyance piping might be a limiting factor for getting all of the rain to one storage. Also, when you need to store large volumes of water — or in cold climates — it can be far more cost effective to have a combination of above-grade and below-grade water storage. Lastly, if you have multiple end-uses (such as both indoor and outdoor rainwater demand), you might choose to locate tanks in two (or more) different locations relative to the end-use requirements.

To help you understand if your gutters and downspouts are a limiting factor, and to help you make your decision of one tank location vs multiple tank locations, we recommend that once you have determined your best storage locations, you contact a local rain gutter specialist (or *eavestrougher* as they are called in Canada). In Chapter 5 we'll explain in more detail why we don't think you should bother trying to do your

Fig. 4.7: *Gutters, slope, existing downspouts, potential tank locations as well as relative distances overlaid on the base map.*

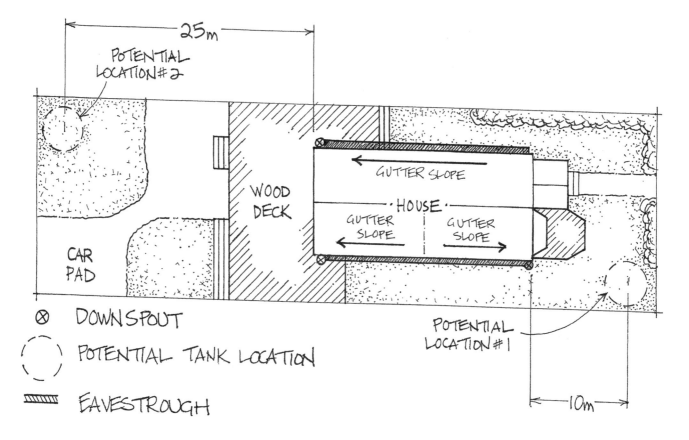

own rain gutter calculations, and why we rarely do them ourselves.

Your rain gutter specialist will look at your existing building (or the building plans), the flow volume limitations of standard rain gutters, and the required slope of the gutters and conveyance piping, along with the site plan showing your proposed tank locations and elevations. They'll use this information to assess the maximum volume that your rain gutters and associated downspouts can handle and help you to confirm if your storage location(s) is/are acceptable.

Based on the feedback of the gutter installation professional, you should be able to make a near-final decision on the placement of your tank(s) and downspouts and/or whether you can reuse existing downspouts.

We call it a *near-final* decision because you still need to evaluate and design your conveyance piping (number 9 on Figure 4.1), which could potentially influence or even change your ideal tank placement. You also still need to consider what is available from manufacturers, in terms of actual tank dimensions and installation requirements, which is covered in more detail in Chapter 6. Depending on how space-constrained you are, this could also influence your final tank placement and layout.

Available Elevation

You've now established the storage locations based on your spatial layout. Your downspouts have been placed, based on a combination of relative placement/proximity to storage and a consideration of maximum water volume capacity.

The next thing you need to evaluate is the vertical plane. The idea here is that water will naturally flow in one direction only — that is, down. You need to make sure that the elevation loss required for water to effectively flow from your gutters through any pre-filtration equipment, then through any conveyance piping and down into your tank, is less than your *available elevation*. Your available elevation is impacted by the tank height, the gutter height, the horizontal distance from the gutters, and any overall site elevation changes. This is illustrated in Figure 4.8.

Fig. 4.8: *Available Elevation is impacted by the tank height, gutter height, horizontal tank distance from the gutters, and overall site elevation changes.*

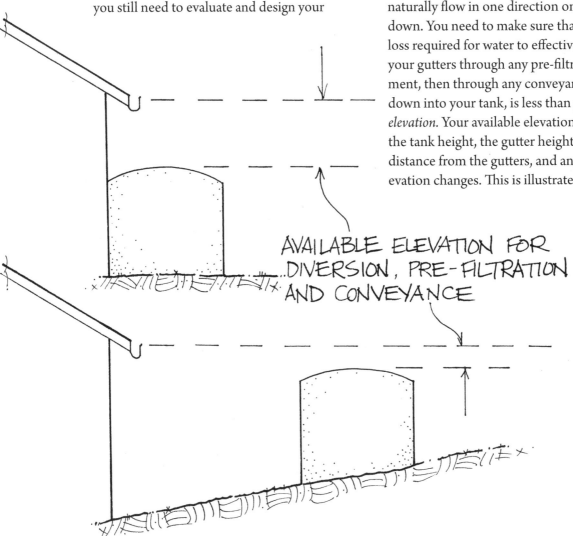

AVAILABLE ELEVATION FOR DIVERSION, PRE-FILTRATION AND CONVEYANCE

The key thing here is that you must make sure that your available elevation is greater than your minimum required vertical distance, which is defined as follows:

Minimum Required Vertical Distance = Vertical Span for Diversion Valve + Vertical Span for Pre-Filtration Equipment + Vertical Span for Fittings + Elevation Change Required for Conveyance

This minimum required vertical distance is illustrated in Figure 4.9.

Note that not all systems will have a diversion valve or use a downspout filter as a pre-filtration strategy, as shown in the image. The minimum required vertical distance is therefore very much related to the required vertical span of the

pre-filtration equipment (and associated fittings) that you select, details of which are covered in Chapter 5. The other consideration you have is the elevation changes that are required for your conveyance piping, also covered in Chapter 5.

If your tank is below grade, you'll likely have no problems exceeding your minimum required vertical distance as you have a lot of available elevation to play with. However, tall above-grade tanks, tanks placed large horizontal distances from the gutters, or sites where ground elevations increase away from the house can pose particular design challenges. You may actually find that you have so little available elevation to work with that it will influence your choice of pre-filtration strategy and/or significantly impact your choice of tank dimensions and/or

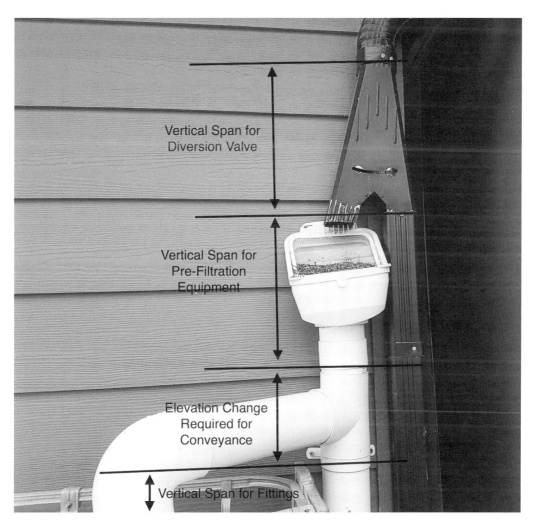

Fig. 4.9: *The minimum required vertical distance is the vertical span required for your diversion valve (if applicable), your pre-filtration equipment, your fittings, and any elevation changes required for your conveyance piping. The Available Elevation (Figure 4.8) must be greater than this minimum required vertical distance.*

influence the design of your conveyance piping. There could even be a scenario where you need to go back and select a new storage location — once again illustrating the iterative nature of RWH system design.

Also, if you are considering raising your tank up onto a tank stand to effectively increase head pressure, you'll want to ensure that your tank stand does not raise your tank above your minimum required vertical distance.

Elevation Plan View Drawing

To help you with your elevation planning and determination of your available elevation, you'll want to draw, or at least start laying out, an elevation plan view, similar to Figure 4.8.

We frequently do our elevation plan view drawings by hand and simply select an appropriate size of paper, an appropriate scale, and once again pull out our architect's scale ruler. Using any relevant information from our site plan (particularly information from the elevation and structures overlays), this drawing is quick to produce and absolutely indispensable going forward.

If you have an elevation constraint and your actual available elevation is small, you'll want to keenly be aware of your equipment selections and conveyance piping design choices and the impact these will have on your minimum required vertical distance.

Frost Protection and Layout

If you are in a climate with a winter cold enough to freeze any outdoor standing water, you have two options:

i.   Design a system following all the typical recommendations for warm climates, but use it only as a seasonal system. Drain and decommission your system every year before freezing occurs and use a diversion valve to direct water away from your tank when the system is decommissioned.

ii. Provide frost protection to your system. This includes protection for all conveyance piping, in addition to water storages.

How you choose to provide frost protection is highly dependent on your context, resources, and intended end uses. We'll provide more details on how you might provide frost protection for your conveyance piping in Chapter 5, but it's necessary to consider how you plan to frost-protect your tank at this stage of the design, as it will have a substantial influence on your preliminary site layout.

Some options include:

- Integrating your water storage into the basement of your home or an existing building that you already plan on heating. We've seen integrated cisterns as well as tank systems installed in basements.
- Provide heating via electricity to an insulated tank. We've seen this done, but don't consider it to be a very resilient design. Consider what might happen if the power goes out in winter time.
- Provide water movement, via an air bubbler or equivalent, to keep water moving and prevent freezing. Once again though, this design depends on a continual supply of power.
- Bury the water storage and all associated conveyance piping below the frost line.

When burying a tank, consider that the farther away it is placed from the downspout, the deeper it will need to be in order to achieve adequate drainage (via slope) for all conveyance piping. The deeper the tank, typically the higher the cost.

If your tank burial depth is driven by the need for frost protection, you can reduce the buried depth by using a horizontal insulation layer installed above the tank, yet still below grade. Think of it as a barrier to preventing frost from migrating downward. See Figure 4.10 for an illustration of this strategy.

Keep in mind that not all insulation is equal and that any insulation (such as expanded polystyrene, EPS) that you place underground needs to be *burial grade.* Also, when you place a horizontal insulation barrier, it will affect the ground moisture; it can become hard to grow anything in the area above the insulation.

Lastly, as below-grade tanks are more expensive than above-grade tanks, using a combination of above- and below-grade storage can be a cost-effective way of meeting year-round demands without breaking the bank. Large amounts of water can be harvested and stored above ground during the warm/rainy seasons to provide for summertime indoor and outdoor use. The below-grade storage needs only to be sized for wintertime demands.

Although we have provided you with a few tips and ideas here, remember that underground infrastructure is complex. We recommend that you do not consider going this route without engaging the advice of an experienced contractor or a professional.

The Importance of Placement

The considerations related to placement of your RWH system components are critical to a successful design.

In many other resources on RWH system design, we've often found that nearly all the focus is placed on the design of individual components. However, we argue that proper placement and consideration of how the individual components function together — as a whole — is critical to ensuring that a system will work successfully and deliver quality water.

In the next chapters, we will narrow our focus down to the individual component considerations, including pros and cons, and design and specification details.

Fig. 4.10: *A horizontal insulation layer can be used to minimize tank burial depth in a cold climate. In this drawing, an EPS board (a rigid foam) is used.*

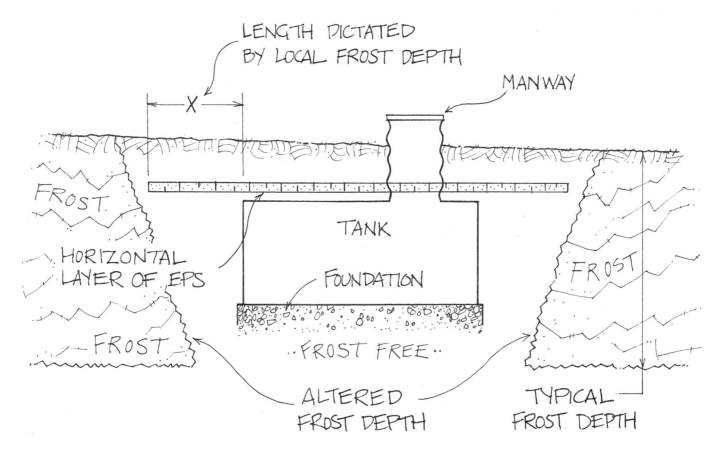

LENGTH DICTATED BY LOCAL FROST DEPTH

MANWAY

X

FROST.

HORIZONTAL LAYER OF EPS

FROST

TANK

FOUNDATION

FROST FREE

FROST

ALTERED FROST DEPTH

TYPICAL FROST DEPTH

Chapter 5

Collection and Pre-Filtration

Your roof, your rain gutters, your downspouts, your pre-filtration equipment, and any horizontal conveyance required to send the water to its storage location make up your collecting and pre-filtration components.

However, in addition to the role these components play in supplying you with quality rainwater, they also play an essential role in protecting your building and the hard structures around your house, such as sidewalks. Improperly installed roofs or incorrect flashing details can cause leaks, rotting of building materials, and, ultimately, failure. Improperly sized gutters and downspouts, or lack of maintenance, may result in overflow during large rain events. Missing, leaking, or misdirected downspouts may result in water accumulation near foundations causing mounding, heaving, and cracking.

Given the severity and cost of repair compared to the relative ease of managing rainwater properly, it's rather astonishing how often we see damage to homes, building walls, foundations, and other hard structures due to improperly managed rainwater runoff.

Therefore, for two very important reasons, (i) the effective delivery of quality rainwater to your storage; and (ii) the protection of your house and associated infrastructure, ensuring high-standard installation and workmanship should be a top priority as you set out to design, build, or make modifications to your rainwater collection components.

Roof

In addition to the choice of material, a number of environmental, design, and maintenance factors all contribute to the quality of water that will run off your roof and be collected by your catchment system.

According to Spinks (2007), there are several simultaneously occurring anti-microbial processes that occur on the roof catchment that increase your water quality:

- **Heat inactivation:** Heat and direct exposure to sunlight are valuable sterilization mechanisms.
- **Desiccation:** Roofs with higher slopes and/or peaks that drain properly will dry out, as compared to flat roofs. This desiccation effect reduces the microbial load.
- **UV radiation:** UV radiation is a powerful disinfectant. Roof surfaces exposed to solar radiation will benefit from this effect.
- **Wind:** Wind can be a cleaning force or a contamination force. Wind may remove larger debris but may also deposit smaller particles.
- **Rain:** Rain will clean the roof; however, the system needs to have appropriate pre-filtration equipment to isolate the first flush of rain and direct it away from the storage components.

There are also environmental factors that *increase* your contamination risk, and therefore reduce your water quality. These include atmospheric deposition (are you nearby to a pollution source?), overhanging tree cover, fecal matter (from avian friends and small mammals), and other debris from trees and plants.

As part of the design process, consider what you can do to enhance the cleaning and sterilization factors while reducing contamination vectors. Once the system is installed, ongoing maintenance practices (such as roof cleaning and the removal of debris and overhead branches) are essential and must be viewed as one part of the overall water-quality plan.

Note, lastly, that as long as you keep livestock and humans off your roof, the majority of bacteria and microorganisms that may make their way into your rain tank will not be pathogenic to you, and will not be of fecal origin. For an interesting, albeit slightly technical, read on coliforms, microbial diversity, and quality of captured rainwater, see the research papers by Evans and by Spinks that are listed in the Resources section.

Selecting Roofing Materials

The biggest decision you typically have to make about your roof is the choice of material. Common choices for North American homes include asphalt (the most common), coated and/or painted metal (steel, aluminum, copper, and zinc), wood shingles, tiles.

Table 5.1 summarizes the regulations and standards that we've seen related to material selection in potable vs non-potable systems.

Unfortunately, there is often very little information available from roofing manufacturers

regarding whether or not their products meet any minimum standards for water quality. At least that is the current situation in North America. As such, you can't often rely on third-party certification to help you select roofing material appropriate for your end-use. You must approach material selection by first reviewing your regulatory context, then by applying a principle-based approach, and lastly by asking good questions of the manufacturer or product supplier. The good news is that this is a pretty intuitive process.

Your best choices for roofing materials are the ones that are nontoxic and the most benign (i.e. unlikely to leach material-based contaminants into your water). You want to select a material or roofing system that is highly efficient in that it collects and conveys water well. Smooth surfaces are preferred, as they are less likely to collect dust and are easiest to clean.

Examples of materials that meet these specifications include factory-coated enameled steel, sheet steel coated with aluminum-zinc alloy (e.g. Galvalume), galvanized metal roofing products, terracotta tiles, concrete tiles, cement tiles, glazed tiles, and slate tiles.

Galvanized metal roofing products (zinc-coated metal) have been discouraged in a lot of RWH design literature because of health concerns. However, we have not been able to find research substantiating this risk. In addition, a galvanized roof would only release significant zinc if the roof were rusty, and even then only a very high concentration of zinc would pose a health risk. We would likely feel comfortable designing systems with zinc roofing material — as long as it is combined with all of the sensible design and maintenance practices outlined in this book.

Pay attention to the following:

• Use only nontoxic sealant or grout with any tile.

Table 5.1: *Typical North American regulatory and standard-stated requirements for roofing materials for potable vs non-potable systems. See the Resources section for the full NSF name references.*

Table 5.1

Potable Systems	Non-Potable Systems
• Recommended that you look for roof systems/materials certified to NSF 151; but given the lack of industry certification, this is typically not an absolute requirement. • If the component is manufactured with a coating, or if a coating (or paint) is applied post installation, coatings must meet NSF 61. • There will be roofing materials explicitly stated as not allowed (lead flashing as an example).	• Regulations for non-potable systems tend to be less prescriptive and focus on the requirement of "delivering water quality appropriate for the end use."

- Ensure that the concrete does not contain fly ash (see the sidebar "Beware of Fly Ash When Purchasing Concrete," in Chapter 6).
- Consider water exposure to the rubber gaskets used on metal roofing fasteners. Go with a metal roof system with concealed fasteners/gaskets, if possible.
- Confirm with the manufacturer that any and all coatings applied meet NSF 61 and/or are nontoxic.

Asphalt shingles are the most popular roofing material for North America homes. However, they are not really a great choice for delivering quality water. They are not recommended where drinking-quality water is desired. If used, an appropriate pre-filtration strategy should be employed (such as a rainhead combined with a first flush diverter). Note also that some manufacturers treat shingles with herbicides to prevent the growth of moss. This could be a huge deterrent to using an asphalt roof as catchment for irrigation.

Wood or cedar shake roofing is not usually approved for potable water use, and water coming off such roofs can end up being be too acidic (and potentially toxic, in the case of cedar) for some plants. These roofing systems are very inefficient and difficult to clean. However, wood and even copper roofing can actually deliver reasonable and even high-quality water with the correct management and construction protocols (P. Coombes, personal communication, March 2018).

There are a few roofing materials that are far less common in home-scale RWH systems, but we'll briefly mention them here for completeness.

- Bitumen or composition roofings are typically used for flat or low-slope roofs. These roofing systems are not recommended (and perhaps not permitted) for drinking water. They are difficult to clean, and ponding in flat spots will degrade water quality.
- EPDM (ethylene propylene diene terpolymer) roof systems are far more common in low-slope and/or commercial applications. Manufacturers may have a product line that has been water-quality approved by third-party certification, such as NSF 61 or NSF P151.
- Lead flashings are not to be used for any RWH system.

Lastly, if you paint your roof, use common sense and stay away from paints that are lead-based.

Even if you do not intend to use your system for drinking and/or bathing, we strongly encourage you to select the highest quality, least toxic, and most benign roofing material available.

Rain Gutters

Rain gutters and downspouts are shown as number 2 and 3 on Figure 4.1. Gutters are essentially metal ditches. They capture runoff from all around the perimeter of your roof and redirect it to your downspouts.

As part of the initial design process, you'll select gutter materials, type, and size. They need to be appropriately sloped toward your downspouts and deep enough that they don't overflow during large rain events.

When it really comes down to water quality, the biggest impact gutters will contribute is tied directly to your maintenance regime (or lack thereof). Make sure you are inspecting your rain gutters at the start, middle, and end of the rainy season. If you notice that there is always debris collected in the gutters (or none at all), you can increase or reduce your maintenance schedule accordingly.

Materials, Styles, and Types

Compared to roofing, there's not nearly as many options when it comes to gutter and downspouts.

Aluminum is the most popular and common choice because it is easy to shape, comes in many colors, and is corrosion resistant. You may find that your rain gutter professional *only* works with aluminum.

Galvanized (zinc-coated metal) is another choice available on the market. It is stronger than aluminum and therefore perhaps a better long-term choice if you live in a climate susceptible to hail.

Aluminum-zinc coated sheet steel (e.g. Galvalume) is harder to find, but is another good choice for systems that require high-quality water.

Copper gutters are not common and we wouldn't recommend them unless they were coated to prevent leaching of copper into your water.

PVC (polyvinyl chloride, also called *vinyl*) gutters and downspouts are a popular choice for DIY homeowners because of their low cost. They are lightweight and come in small sections for easy transport and installation. However, we are not big fans of PVC gutters. Overall, given the important role that gutters play in protecting your home and infrastructure from water damage (in addition to the role of transporting water to your tank), we feel that they are a poor-quality choice, with a short lifespan. We also don't like that there is no seamless option (discussed below). As such, we never recommend them on any project we are involved with.

If you are looking for drinking-quality water, there are several other concerns about PVC gutters. First off, there could be lead residue in the PVC material itself, as lead is sometimes used in the manufacturing process. Secondly, PVC has been shown to break down in sunlight and form carcinogens that could leach into water. If you really want to use PVC gutters and downspouts, make sure that they are coated (to protect from UV degradation) and that lead was not used in the manufacturing process. See the sidebar, "Should You Use PVC?" below for an expanded discussion on the use of PVC.

Gutter Shape

The shape of your gutter affects how water is conveyed from the roof to the conveyance piping. There are two common shapes, referred to as *K-style* and *half-round*.

K-style has a flat bottom, and the trough holds more water per unit length of gutter than the round gutter. In addition, they are typically stronger than half-round gutters. Because of this, K-style gutters can be built in longer lengths; this means they can be installed with fewer seams, making them less susceptible to leaks.

Water will flow faster through a half-round gutter than it will a K-style gutter because of

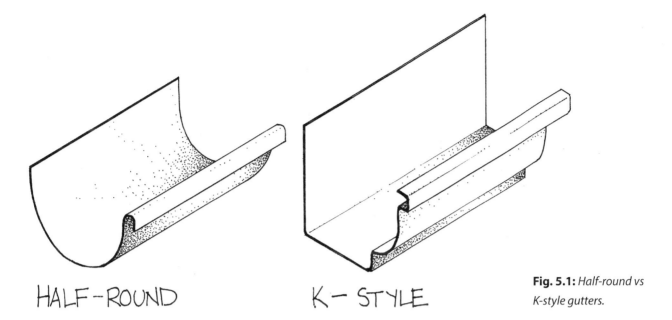

HALF-ROUND   K-STYLE

Fig. 5.1: *Half-round vs K-style gutters.*

the smaller cross-sectional area. Because of this, half-round gutters should self-clean better than K-style because, in theory anyway, they should more readily allow debris to travel through them. However, you certainly shouldn't rely on this as a replacement for ongoing maintenance and inspection.

Seamless vs Seamed

The gutter seam is the place where two sections of gutter are joined together.

If you purchase off-the-shelf gutters from your local hardware store, you'll typically end up with 3 m (10 ft) lengths. As roof lines are typically much longer that this, you'll have to join several pieces together using exterior-grade caulking. The seams are the places in your gutter system most likely to develop leaks over time, and installers make up for this shortfall by installing seamed gutters at a steeper slope (thus increasing ongoing drainage toward the downspouts).

Seamless gutters are gutters made specifically for your building using specialized equipment. Seamless gutter installers show up onsite with massive rolls of metal in their trailer, and their

machine draws metal from this roll, shapes it through rollers, and extrudes gutter of almost any length. Seamless gutters are far less likely to leak, and they can be installed on a shallower slope (typically 1/10 of the slope required for seamed gutters).

You can't build and install seamless gutters on a do-it-yourself basis, and we believe that hiring a seamless gutter installer is one of those times you'll get good-to-excellent value; it's usually just not worth trying to do it yourself with prefab lengths. Getting a gutter professional involved not only means a better-quality product installed on your home, but it also means that you'll benefit from their help with your gutter sizing, volume calculations, flow limitations, and placement of your downspouts. Keep in mind, too, that poorly or improperly designed gutters are likely to be the largest source of water loss in a RWH system.

Sizing Gutters

We'll reiterate again that we think that the most effective use of your time and energy invested into the design of your RWH system should go first into the upfront planning and feasibility (Chapter 3) and then the site planning (Chapter 4). From there, invite a gutter installation professional to your site and present them with your site plan, elevation plan drawings, and proposed tank location(s).

Based on the maximum rainfall intensity for your area, along with a site evaluation, this information will provide everything they need to be able to quickly and effectively tell you:

- What size gutters you need
- Where downspouts should go, and what size
- How to slope the rain gutters
- What horizontal conveyance might be required

If you really want to size and install your own gutters see *Rainwater Harvesting: System Planning,* Mechell et al. (2010), in the Resources section of this book, or visit Brad Lancaster's website: www.harvestingrainwater.com.

Splitting Up Tank Volumes Based on Downspouts

If you are using multiple tanks in multiple locations, you can now proportionally split up the water volume that should be stored at each tank location based on the proportional volume contribution of each downspout. If downspout A is capturing 60% of your roof runoff, then the tank associated with downspout A should be designated 60% of the total required storage capacity that you calculated in your feasibility analysis, from Chapter 3. See Example 5.1 for a sample calculation of this using a site plan.

Example 5.1: Splitting up tank volumes based on relative downspout catchment area.

A RWH system designer has selected three downspout locations, and marked these locations on her 1:100 metric scale site plan,* as shown in the figure below. Based on slope, she determines which roof sections are draining to which downspouts and marks them out accordingly.

Using a 1:100-scale architect's ruler (metric), she measures,** directly from the drawing, the dimension of each roof section: roof section A is 13 × 5 meters, roof section B is (11 × 5) + (2.6 × 2.4) meters, and roof section C is 7 × 2.6 meters. The associated catchment area for each of these roof sections are therefore: 65 m² (45% of total catchment), 61.2 m² (42.4% of total catchment) and 18.2 m² (12.6% of total catchment).

She has three tank locations planned at the base of each downspout. From her feasibility analysis, she knows she wants a total storage capacity of 5,000 liters (1,320 gals) and therefore plans to split that volume between her three locations as follows:

Tank A, Ideal Tank Volume = 5,000 liters × 45% = 2,250 liters

Tank B, Ideal Tank Volume = 5,000 liters × 42.4 % = 2,120 liters

Tank C, Ideal Tank Volume = 5,000 liters × 12.6% = 630 liters

*The closest imperial scale is ⅛" = 1'-0," which is 1:96. You can therefore use an imperial scale ruler and should be able to get close to the following measurements (converted from meters): A: 43 × 16 ft, B: (36 × 16 ft) + (9 × 8 ft), C: 23 × 9 ft.

**If you have a 1:100 architect's scale ruler, and are comfortable in metric, see if you can prove this to yourself.

Aerial view of roof areas and downspouts A, B, and C at a 1:100 scale. This is the authors' home in Calgary, Alberta.

Diversion Valves

A diversion valve is shown as number 4 on Figure 4.1.

This valve simply has a flap that can manually be switched to divert water one direction to another. Here are some reasons you might include one in your design:

- In any system, but particularly if you anticipate having problems with debris on your roof and in your gutters, you could use it as a way of re-directing roof wash water or accumulated debris from gutters when doing regular maintenance and cleaning.
- In climates where the rain comes in one season only (or mostly so), it could be used as a manual "first flush system" by manually diverting the first rain at the start of the rainy season.
- In seasonal cold-climate systems, you can use it to divert water away from the filtration and tank during months that the system is decommissioned.

As always, ensure that the water from your diversion valve is directed at least 3 m (10 ft) from any home or tank foundation.

Pre-Filtration

The goal here is to reduce the amount of debris in rainwater storage — thus reducing the potential for contamination. Decomposing leaves and other organic matter will create foul-smelling rainwater if they enter into your storage.

Rainwater harvesting equipment distributors sell pre-filtration equipment as off-the-shelf components. Your job is to select the correct component, or combination of components for your particular context, environmental conditions, and required end-use water quality.

Because nearly all pre-filtration equipment is based on the gravity flow of water, many components require a certain vertical span. If your Available Elevation is small (see Chapter 4), you

may make your equipment or strategy selections based on keeping your minimum required vertical distance below the Available Elevation. Alternatively, if you are considering placing your tank on a tank stand, you'll want to select your pre-filtration strategy first, and subsequently determine your minimum required vertical distance, such that you can raise your tank but still have enough space for your pre-filtration equipment.

Note also that maintenance is absolutely crucial for successful performance. A lack of maintenance of pre-filtration can reverse the intended effect; instead of improving the water quality, your pre-filtration will become a liability and ultimately substantially degrade the quality of your water. This requirement for maintenance must be a priority consideration in the selection of your equipment, the placement of your equipment for ease of inspection, and in setting up a long-term maintenance schedule.

Below, we provide an overview of the most common types of pre-filtration equipment available on the market.

Screens

Screens are arguably the most important pre-filtration strategy, and you'll absolutely want to include some sort of screen — better yet, several screens in several locations. There are four main styles/locations: gutter screens, downspout filters, centrifugal filters, and in-tank filters.

In addition to acting as a sieve and separating and removing larger particles from the water, a good screen is also required to prevent insects and rodents from getting into your water storage.

Gutter Screens

There are numerous commercial gutter screens on the market. The idea is that these screens, placed overtop of the gutter trough, reduce the amount of debris that gets collected into the

RWH system. However, we generally avoid them and recommend that you do too.

Most models, once installed, are not easy to remove. This, in of itself, creates additional difficulty when it comes to the maintenance and inspection RWH systems require.

The screens prevent sun from getting in and can stop or slow the desiccation of the trough, allowing water to stagnate. As previously mentioned, the UV disinfection action on the roof and gutters represents an important step in the incidental treatment train in a RWH system.

Screens tend to also slow the flow of water and the subsequent scouring, or self-cleaning, effect that occurs in large rain events, increasing the potential for debris accumulation in the gutter. In addition, stagnating water and debris tends to cause corrosion and holes in the trough, which means the gutter may need to be replaced prematurely.

One of our most reputable and experienced contractors had quite a few negative things to say regarding gutter screens — based on his 30 years of experience. He felt that these screens tend to "move the problem up by four inches." Having a ledge on the trough means that larger debris that sits on the trough screen rots close to the edge of the roof and can compromise the decking of the roof itself. In addition, this debris can plug off the screen holes and can create ice damming along the roof edge. Tellingly, he said: "I have removed far more than I have put in."

Downspout Filters

A downspout filter is shown as number 6 in Figure 4.1. It is also commonly called a *rainhead*. A downspout filter is installed inline of the downspout in the vertical orientation. The screen at the top of the unit allows rain through while holding debris back.

Some models have a screen at a 45° angle, allowing for filtered debris to be flushed off the screen during larger rainflow events. Although manufacturers might promote them

Fig. 5.2: *Downspout filter on an off-grid cabin.*

as self-cleaning, we have found that in practice these screens still require manual inspection and debris removal at an interval dependent on your site conditions and choice of roof material.

When comparing different models, look for a screen that is stainless steel (corrosion resistant) with a mesh size less than 1mm; it should be removable (for easy maintenance). Also, if you place the downspout filter in a sunny location, and if you use a model with an exposed screen (as compared to enclosed), your screen — and water quality — will benefit from the UV and desiccation effects of sunlight.

 Advantages:

- Simple, easy to install, inexpensive, and relatively effective.
- Prevents vermin and insects from entering your system.
- Quick and easy to clean.

- Comes in 80 mm (3 in) and 100 mm (4 in) variants and works well in low and high flow rates.

 Disadvantages:

- Susceptible to freeze/thaw issues.
- Requires regular manual maintenance and cleaning.
- Requires vertical span.

In cold climates, a downspout filter can be used effectively if the system is decommissioned in the winter and a diverter valve redirects water away during that time. This configuration option is shown in Figure 4.1.

Rain tank Inlet/Basket Strainers

A basket strainer is simply a screened basket placed at the tank inlet. We highly recommend including one of these in your design.

Fig. 5.3:

A basket filter in a tank.

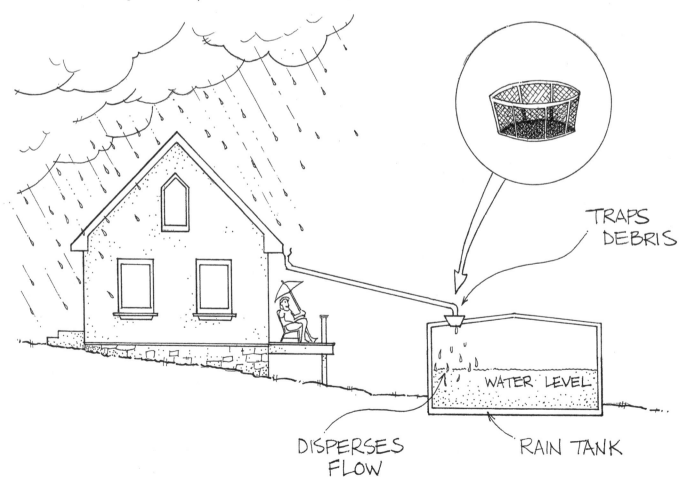

When installing a basket filter, be sure that the tank overflow (i.e. max water level) is lower than the bottom of the filter. Otherwise you risk turning your basket filter into a "tea bag" when the water level is high.

 Advantages:

- Inexpensive and very effective at removing debris from the rainwater stream.
- Diffuses rainwater energy as it passes through the basket, reducing turbulence in the tank (works as a *quiet inlet*).
- Requires the least amount of vertical span and therefore a good option if Available Elevation is very small.

 Disadvantages:

- Any debris caught in the basket will stay there until you remove it (a colleague once found a dead rat in a basket filter while doing an inspection).
- Requires regular inspection and maintenance; it is not as easy to visually inspect as a downspout filter is.

Centrifugal Filters

As the name suggests, these filters use centrifugal force and work on the principle that water is heavier than debris. Consider that the centrifugal force (and hence the filtration capacity) will be a function of flow rate. As such, a manufacturer should provide an ideal operational range in terms of flow rate. There are versions of this filter that are installed inline with the downspout, and some that are installed in the tank itself. If you operate the unit outside of its operating parameters, you may not get the same level of performance and filtration.

If you live in a climate where the rain is predominantly light (i.e. frequent pitter-patter), this type of filter may not be effective. Check with the manufacturer's specifications.

 Advantages:

- Very effective at removing debris (within its operational range).
- Usually self-cleaning and requires less maintenance that other filter types.
- Water passing through the filter is oxygen-enriched, which can improve water quality.

 Disadvantages:

- More difficult to size properly and works best if rain intensity is relatively steady.
- More expensive.
- May require more vertical space than other pre-filtration units.
- May not be effective in low-intensity rain events.

In general, given its added cost and complexity, we think that centrifugal filtration may be overkill for most home-scale RWH systems. If you follow the good design and sensible maintenance practices outlined in this book and support the inherent RWH system treatment train that will result, a downspout filter combined with a rain tank inlet/basket strainer is usually a sufficient and simpler alternative.

Fig. 5.4: *WISY cyclone (centrifugal) filter.*
Photo credit: WISY (www.wisy.de)

First Flush Devices

A first flush device (Figure 4.1, number 7) works on the idea that the first flush of runoff water from the roof may be the dirtiest water. By collecting this "first flush" water separately, then disposing of it, you can reduce the amount of metals and pollutants that are sent to your storage tank. According to Coombes (2015), these devices can remove 11%–94% of dissolved solids and 62%–97% of suspended solids from the first flush of runoff. Note, however, that that proper design of your storage (for sedimentation) and the inclusion of screens might achieve similar ends.

First flush devices come in a few different configurations, but the general idea is that when it rains, the first runoff fills a vertical pipe of a pre-determined volume. Water only starts to flow into the tank once this column is full. A small drain port (Figure 4.1, number 8) allows for the slow, eventual release of the diverted water.

It is the regular maintenance of the drainage port that determines the success (or the total failure) of this device. As small debris slowly accumulates at the bottom of the vertical column over time, it plugs off the drainage port. As a result, this contaminated water can go septic, and rather than improve water quality, it has a detrimental impact. Because of this maintenance reality, there are numerous industry professionals who don't like using them or installing them in their systems.

Inspection and maintenance is absolutely critical for first flush devices. If you know from the onset that you (or the homeowner) may not make maintenance a priority, then you may be better off without a first flush device as part of your pre-filtration strategy.

In addition, if the end-use does not require high-quality water, a first flush device may not be worth the increased configuration complexity and the increased long-term commitment to maintenance.

There's also a trade-off with a first flush diverter in that it will reduce the total amount of your captured rainwater. Given that in most places rain tends to come in numerous small events (vs one or two large events), oversizing the first flush diverter can have a significant negative impact on water volume captured.

With all of the considerations above, we would tend to include one in a design if we thought that the surrounding pollution loads were high (near an industrial area, higher-pollution area, etc.) and then put operational measures in place, such as simple calendar reminders, to ensure that its inspection is not neglected. Otherwise, we'd likely select a downspout filter and tank/basket filter combination as our pre-filtration strategy.

When sizing a first flush diverter, we recommend that your diverted volume be no greater than 20 liters (5.3 gal). Beyond that, you are reducing your water yield for relatively little gain in water quality.

Commercial first flush devices usually come in kit form, including the seating ball, seals, and the drainage port. You simply add the correct length and diameter of pipe (typically PVC) to accommodate the volume that you want to divert. To make your piping length and diameter selection easy, we have included common pipe diameters, lengths, and their associated volume holding capacity in Table 5.2.

From Table 5.2, to divert ~20 liters (5.3 gal) of rain, you could select 2.4 meters of 100 mm diameter pipe (8 ft of 4") or 1.2 meters of 150 mm diameter pipe (4 ft of 6").

If your conveyance pipe is 80 mm (3") or 100 mm (4") in diameter, commercial plumbing distributors carry common reducers to adapt the pipe to any number of diameters.

Other Types of Pre-Filtration

Not all equipment-types and strategies are discussed here because of the numerous manufacturer-specific variations and options. Therefore, before making any selection, you'll want to be checking with your local supplier to see what is available. You'll also want to review all of the manufacturers' specifications.

In general, you'll find that many suppliers carry combination units that act as both first flush and downspout filters, as shown in Figure 5.5.

There's also whole category of specialty filtering devices for both above- and below-grade tanks. These devices are very efficient but quite a bit more expensive and relatively complex. Many models claim to be entirely self-cleaning, many require power and some may only work with manufacturer-specified tanks.

Conveyance Piping

Unless your tank inlet is located directly beneath your downspout and/or directly inline with your pre-filtration equipment, you are going to need some pipe to redirect water appropriately. This pipe is called *conveyance piping* and is indicated as number 9 in Figure 4.1.

Depending on the configuration of your piping, and the intent for it to drain (or not drain) water, we call it either *dry conveyance* or *wet conveyance*.

In a dry conveyance system, you design your piping such that all water drains out of the pipes in between rain events. To accomplish this, the conveyance piping is installed with a slight slope toward the tank.

In a wet conveyance system, the piping does not drain, and water is trapped in the lowest point(s) between rain events. When a rain event occurs, you rely on the elevation differential between the inlet of your conveyance piping and the outlet of the conveyance piping such

Fig. 5.5: *Combination downspout filter and first flush diverter (2-in-1 unit).*

Table 5.2

Pipe Length	Volume, liters [gal]*		
m [ft]	Ø100 mm [4 in]	Ø150 mm [6 in]	Ø200 mm [8 in]
0.3 [1]	2.6 [0.7]	5.7 [1.5]	9.8 [2.6]
0.61 [2]	5.3 [1.4]	11.4 [3]	**19.7 [5.2]**
0.91 [3]	7.9 [2.1]	17 [4.5]	29.5 [7.8]
1.22 [4]	10.6 [2.8]	**22.7 [6]**	39.4 [10.4]
1.52 [5]	13.2 [3.5]	28.4 [7.5]	49.2 [13]
1.83 [6]	15.9 [4.2]	34.1 [9]	59.1 [15.6]
2.13 [7]	18.5 [4.9]	39.7 [10.5]	68.9 [18.2]
2.44 [8]	**21.2 [5.6]**	45.4 [12]	78.7 [20.8]

* Beyond ~20 liters [5.3 gal] you get a diminshing return. We've bolded the pipe length and diameter combination that provides this volume.

Table 5.2: *Common pipe diameters, lengths, and the associated volume holding capacity.*

that your rainwater head pressure can drive the standing water through the system.

Figure 5.7 illustrates the configuration difference between these two conveyance piping designs.

Wet vs Dry Conveyance

Consider the advantages and disadvantages of wet conveyance versus dry conveyance stated in Table 5.3.

Given the choice between the two, we prefer dry conveyance systems. Better water quality results, and fewer maintenance issues arise, particularly in our cold climate.

However, we do occasionally find excellent applications for wet conveyance systems. One is when you need to move water underground, or over longer distances, for instance to the other side of a road or pathway.

Figure 5.8 shows one such example. This client had a garage with a driveway and an above-grade tank was located on the opposite side of the building from the downspout. The wet conveyance system moves water through buried piping under the driveway. The elevation differential between the inlet to the conveyance piping and the tank is shown by height "A" on the drawing. As part of the design process, we had to ensure that this resulted in enough head pressure to convey the water all the way into the tank.

In the next sections, we will discuss how to design and select the correct pipe size for both dry and wet conveyance systems. However, note that a nominal pipe size of 80 mm (3 in) PVC schedule 40 potable-grade pipe will most likely be your choice of material and size. That size is about right for more than 90% of the home-scale installations that we've seen. It's easy to get at any local hardware or plumbing supply, and most pre-filtration equipment will be designed to connect easily with 80 mm (3 in) PVC fittings.

Fig. 5.7:

Configuration difference between wet vs dry conveyance systems.

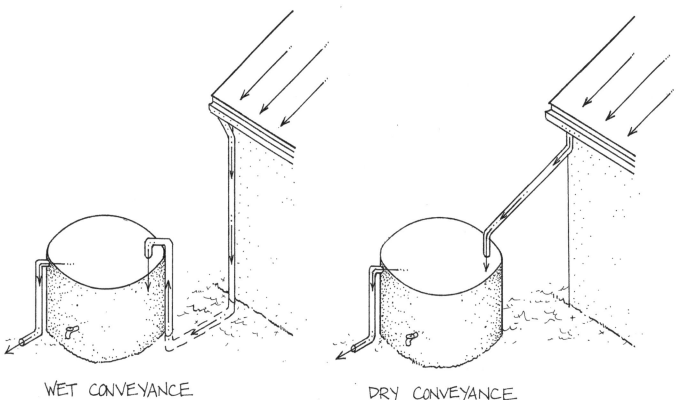

WET CONVEYANCE   DRY CONVEYANCE

Table 5.3: Wet vs Dry Conveyance

Wet Conveyance	Dry Conveyance
Advantages • Piping can be configured such that there is less of an overhead hazard (or less of a tripping hazard if buried/covered). • Can move water a far distance without being concerned about slope and elevation losses required for drainage.	**Advantages** • Less risk that conveyance piping degrades water quality. • Simpler to size. • More appropriate in cold climates.
Disadvantages • Should be emptied in between rain events if no downspout filter in place. If downspout filter is in place, should be emptied after long dry periods. • Pipe breakage can result if not emptied prior to freezing temperatures. • Increased risk of contamination, mosquito infestation, algae growth, etc., and subsequent increased risk of contaminating your storage tank. • The lowest point in the pipe will tend to get clogged with dirt and gunk and could be difficult to clean out.	**Disadvantages** • Pipe runs can be spatially inconvenient, block access ways, or reduce aesthetics. • Requires careful consideration of slope and drainage. • Can significantly increase the minimum required vertical distance when tank is located a far distance from the downspout.

Fig. 5.8:
Example of a wet conveyance system to move water under driveway access. This system includes the ability for gravity drainage of the conveyance piping at its lowest point.

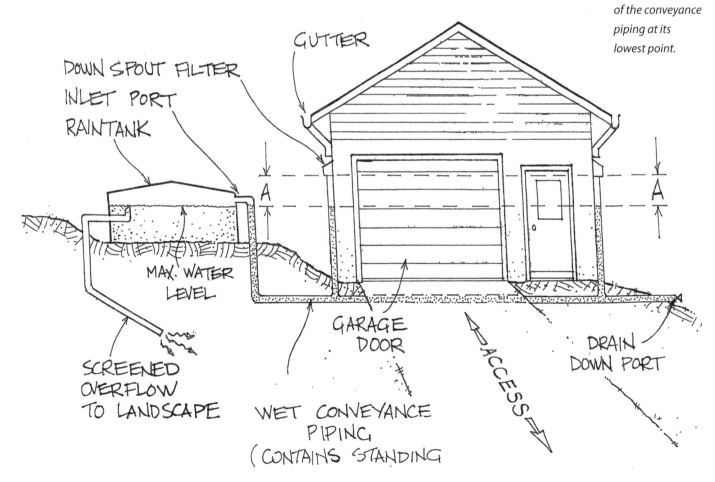

Should You Use PVC?

First off, be wary of *old* PVC (polyvinyl chloride) piping or PVC piping that is not certified to *potable grade.* Old PVC and non-potable rated PVC may have been extruded in lead, or lead may have been used as a stabilizer during the manufacturing process. If the PVC is not certified, you may be able to check with the supplier or manufacturer directly about the manufacturing process, although we have found in practice this is difficult to do.

Several green building organizations are recommending against the use of PVC pipe. On their website, Greenpeace states that the production of PVC uses copious amounts of chlorine and releases dioxin, a toxic chemical. It is also difficult to reuse and as such ends up in landfills.

While there are good alternatives to PVC for gutters and downspouts, unfortunately, the alternatives to PVC for your conveyance piping are limited and few. High-density polyethylene (HDPE) is often recommended, but we've found it difficult to source in 80 mm (3 in). It also requires specialized equipment for welding/melting section and fittings; on the plus side, these joints are far more leak-proof than PVC glued joints.

It can't hurt to ask your local plumbing supply if they carry alternatives to PVC, even though the answer is likely to be no. But the more people ask for it, the better the chance that suppliers will start carrying more options.

The Resources section contains an online link to the "PVC-free Pipe Purchasers' Report," published by The Healthy Building Network.

Always try to place your PVC piping out of direct sunlight and/or provide UV protection with a light-colored PVC-compatible paint. From a maintenance perspective, replace any PVC that is old or that is starting to degrade.

Maximum Design Rainfall Intensity

The design of both wet and dry conveyance piping (as well as the design of your gutters and downspouts) requires you to define the largest short duration surge of water that you want your system to be able to handle in a heavy rain event. We call this short duration surge the *maximum design rainfall intensity.* Using too small of a number for your maximum design rainfall intensity can result in undersized gutters, downspouts, and piping. During a surge, water may overflow, resulting in losses in the volume of captured rain and could also result in damage to infrastructure and foundations (especially if the maximum surge volume is exceeded regularly). Using too large of a number means that you'll specify larger piping than you actually need, resulting in a higher cost for your system (for instance 100 mm [4 in] is quite a bit more expensive than 80 mm [3 in] piping).

In much literature on the design of RWH systems, in addition to the IAPMO Uniform Plumbing Code (the plumbing code), we've seen it suggested that the maximum design rainfall intensity for gutters and associated piping be set to the *60-minute duration, 100-year return.* The 60-minute duration, 100-year return is the largest rainfall event in the last 100 years averaged over the course of a one-hour time step for your particular climate.

However, consider the last very heavy rainfall that you experienced. Very likely there was a peak period of intensity that lasted only several minutes. Therefore, if you wanted better assurance that your system would perform well (i.e. not overflow) during intense rain events, you'd be better off to use something like the *5-minute duration, 25-year return.*

Compared to Climate Normals, rainfall intensity data can be a little tricky to find. There are three places that we can recommend that you look:

1. **Rainfall Intensity-Duration Frequency (IDF) Charts or the Rainfall Frequency Atlas** Rainfall intensity data is often published by a government agency. Accessing charts and data is usually free, and they can be found quite easily by performing an online search. However, how the data is presented can vary quite significantly (maps vs charts vs tabular format); it can be difficult to read; and we've found that you may have to do some digging to find the one piece of data you are after. Pay careful attention to units when you are pulling data and numbers. For instance, the Canadian IDF charts display the intensity as mm/hour, no matter the time step; while the US data is presented as inches, dependent on the time step chosen (5 min, 10 min, 60 min and so on). See the Resources section for website urls and instructions for the Canadian IDF charts, the US Rainfall Frequency Atlas, and the Australian Intensity Frequency Data.

2. **Our Website** Knowing that intensity data can be difficult to find, we've pulled together a table of the 60-minute, 100-year return event for the major Canadian and US cities and included it as a free download on our website. However, should you choose to use the number from our tables, we caution you to cross check with another method. Visit www.essentialrwh.com to download.

3. **Plumbing Codes** If you can get your hands on a copy of your local plumbing code, you'll find data tables for the 60-minute, 100-year return by city in one of the Appendices in a simple tabular format. Unfortunately, plumbing codes are typically quite expensive to purchase. An exception to this the California Plumbing Code. Appendix D: Sizing Storm Water Drainage Systems is available free online, and it contains data for all major US cities. The *Uniform Plumbing Code* (IAPMO) is also available in an ebook version, see the Resources section of this book.

Sizing Dry Conveyance

Sizing your dry conveyance piping is a little bit of a two-part process. You'll first want to determine the elevation losses required for your dry conveyance, driven by the requirement for the piping to properly drain. Then you'll check to make sure that your conveyance piping is large enough to handle your maximum design rainfall intensity.

Elevation Losses For Dry Conveyance

The recommended best practice for the proper drainage of dry conveyance piping is that it is sloped at a 1% grade. Note that in Figure 4.1, the horizontal conveyance piping indicated as number 9 is very short, and therefore there is virtually no required vertical change in elevation. However, when the tank is placed farther away, the vertical drop required can become significant.

To calculate the required vertical drop for dry horizontal conveyance piping, do the following:

- Take your site plan, from Chapter 4, and measure the horizontal distance between your downspout and your tank. This should be very easy to do using your architect's scale ruler at the correct scale indicated by your site plan.

- Multiply this horizontal distance by 1% to determine the required change in vertical distance. For instance, if you measured 10 m (32.8 feet) you'll get 0.1 m, or 10 cm (3.9 inches).

- Use that vertical distance as the elevation loss required for your dry horizontal conveyance piping. Combine this number with the vertical spans required for all of your pre-filtration equipment.

From Chapter 4:

Minimum Required Vertical Distance =
Vertical Span for Diversion Valve + Vertical Span for Pre-Filtration
Equipment + Vertical Span for Fittings + Elevation Change
Required for Conveyance

- Check your elevation plan view drawing and make sure that your actual Available Elevation is greater than your minimum required vertical distance.

Full Pipe Flow vs Circular Flow

During a rain event, water will flow off your roof, be captured by your gutters, and then redirected via gravity to your downspouts. This gravity drainage spirals the water down the inside wall

of the piping, forming a vortex which subsequently draws in a substantial volume of air, as shown in Figure 5.9.

During a rain event in a dry conveyance system, the downspouts and conveyance piping are only ever partially full of water, and the system benefits from this circular flow pattern because it creates a centrifugal force, which promotes particle separation and pre-filtering of the water. In a wet conveyance system, there will be a circular flow pattern in your downspout until it hits the standing water column left over from the last rain event (see Figure 5.8). The *wet* part of the conveyance piping will be full pipe flow (also called *jet flow*).

The methodology presented below for sizing dry conveyance piping is based on the assumption that the pipes are 100% full of water. This assumption is *fundamentally incorrect* for dry

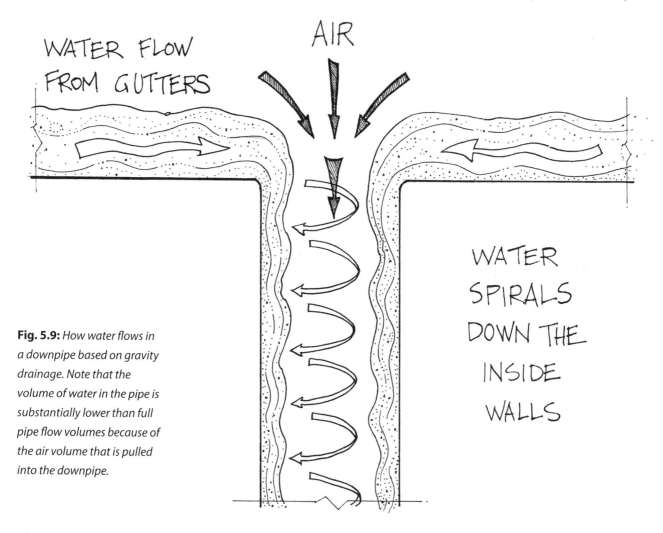

Fig. 5.9: *How water flows in a downpipe based on gravity drainage. Note that the volume of water in the pipe is substantially lower than full pipe flow volumes because of the air volume that is pulled into the downpipe.*

WATER FLOW FROM GUTTERS

AIR

WATER SPIRALS DOWN THE INSIDE WALLS

conveyance. The reason we stick with this incorrect jet-flow assumption for dry conveyance is to be consistent with the plumbing code and North American industry practice — despite the inherent flaw. However, we will point out the errors that result, but also discuss why everything ends up okay in the end.

The reality is that in circular flow situations (i.e. real-life dry conveyance), you will have far less water flowing through your pipes than what would be estimated using full pipe flow assumptions — and it will flow through much slower. As such, the calculations end up being rather conservative, which means that when you stick with the full pipe flow assumptions, you may end up with larger-diameter piping than you actually need.

As discussed in the previous section, the plumbing code recommends that you use the 60-minute, 100-year return as your maximum design rainfall intensity, whereas in reality, it is likely the 5-minute peak intensity (or less) that is the important volume for your gutters to be able to handle. Note that the 5-minute peak intensity could easily be a magnitude level higher than the 60-minute event. Therefore, using the 60-minute event does reduce the conservatism of the jet-flow assumption. In the end, it's a little crude (we'll admit that two really poor assumptions that balance each other out is not good engineering), but it is the common recommended practice. It works out in the end,

and there doesn't appear to a simple alternative that is widely promoted (at least we haven't yet found it).

If you are going to use the methodology presented below for sizing your dry conveyance piping, we recommend that you use the 60-minute, 100-year return as your maximum design rainfall intensity. If you want to be more exacting with your dry conveyance sizing, seek out a resource that provides calculations and guidance based on circular fluid flow (*not* full pipe flow) and use the 5-minute duration, 25-year return as your maximum design rainfall intensity in all calculations. We'll keep our website up-to-date with these resources as we find them as well.

Lastly, there are off-the-shelf *siphonic* downspout devices available for home-scale RWH systems. These devices convert the circular flow pattern into full-pipe (jet) flow. However, we do not recommend these for home-scale systems.

Max Capacity For Dry Conveyance Based on Full Pipe Flow

You now need to confirm that your dry conveyance piping is large enough to handle your maximum design rainfall intensity, which, if you are using this methodology, you'll define as the 60-minute, 100-year return for your climate.

From there, refer to Table 5.4, which provides the maximum catchment in m^2 (ft^2) that your conveyance piping can handle.

	Nominal Pipe Size mm [in]	60-minute duration, 100-year return				
		25 mm/hr [1 in/hr]	50 mm/hr [2 in/hr]	75 mm/hr [3 in/hr]	100 mm/hr [4 in/hr]	125 mm/hr [5 in/hr]
1% Slope	80 [3]	305 [3,288]	153 [1,644]	102 [1,096]	76 [822]	61 [657]
	100 [4]	700 [7,520]	350 [3,760]	233 [2,506]	175 [1,880]	140 [1,504]
2% Slope	80 [3]	431 [4,640]	216 [2,320]	144 [1,546]	108 [1,160]	86 [928]
	100 [4]	985 [10,600]	492 [5,300]	328 [3,533]	246 [2,650]	197 [2,120]

Table 5.4: *The maximum roof catchment area in m^2 (ft^2) for pipe diameter, slope, based on the maximum design rainfall intensity being set to the 60-minute, 100-year return. The numbers in this table are based on the assumption of full pipe flow.* Adapted from IAPMO. 2018. Uniform Plumbing Code. Table 1103.2. International Association of Plumbing and Mechanical Codes. Los Angeles, CA.

For example, assuming that you were planning on using 80 mm (3 in) pipe at a 1% slope, and your 60-minute, 100-year return was 45 mm/hr (1.8 in/hr), the maximum catchment area that you could expect your conveyance piping to handle is approximately 153 m² (1,644 ft²). If your actual catchment was 200 m² (2,153 ft²), you'd be advised to increase to a 100 mm (4 in) pipe.

Sizing Wet Conveyance

Sizing wet conveyance is a little more involved. It's not terribly complicated, but you will have to pull out your calculator or your spreadsheet and your elevation plan view drawing.

Head Pressure

The first thing you'll want to do is to take out your elevation plan view drawing. If it is not done already, identify the elevation/height of your gutter. Properly indicate the vertical span for your diversion valve (if one is included in your design) and the vertical span of your pre-filtration equipment.

Next draw out the configuration of your conveyance piping, taking care to identify all of the fittings you'll need (for instance 45° or 90° elbows).

Next you'll determine the elevation differential between the inlet of your conveyance piping (which is also the outlet of the pre-filtration equipment) and the inlet port of the tank. You'll be able to measure this directly off your elevation plan view drawing using your architect's scale ruler if you've kept everything to scale. In Figure 5.8 you can see the inlet port on the side of the tank and the resulting elevation differential shown as "A."

The elevation differential, in meters (feet) is equal to your head pressure in meters of water (feet of water).

For instance, if you have an elevation differential of 3 meters (10 feet), your head pressure is 3 meters of water (10 feet of water).

You can see that the temptation might be to place the pre-filtration equipment as high as possible in order to increase the head pressure. However, we caution you to always consider that this equipment must be easily accessible for ongoing inspection and maintenance. Therefore, your decision to go with a wet conveyance piping might also impact your choice (and the resulting vertical span and placement) of your pre-filtration components.

Max Capacity For Wet Conveyance

Same as with dry conveyance, you'll need to make sure that your piping diameter is large enough to handle your maximum design rainfall intensity.

First, you'll convert your maximum design rainfall intensity to a maximum flow rate in m³/s (gpm). If you are using the 60-minute, 100-year return (as recommended by the plumbing code) you can do this by using the equation below:

Metric

$$\text{Max Flow Rate} \left(\frac{m^3}{s} \right) =$$
$$\text{catchment area } (m^2) \times \text{roof efficiency} \times \text{60-minute}$$
$$\text{return } \frac{mm}{hr} \times \frac{1}{3.6 \times 10^6} \frac{m}{mm} \frac{hr}{sec}$$

US Customary

$$\text{Max Flow Rate (gpm)} =$$
$$\text{catchment area } (ft^2) \times \text{roof efficiency} \times \text{60-minute}$$
$$\text{return } \frac{in}{hr} \times 0.0104$$

If you are using the 5-minute, 25-year return (or a different time step interval) pay careful attention to your units as you convert this number to a flow rate in m³/s (gpm).

Equivalent Length of Fittings

To simplify the calculation of total frictional losses (done in the next step), you'll convert all of the fittings in your conveyance piping to an equivalent length of pipe.

The equivalent pipe length of common fittings is shown in Table 5.5.

Table 5.5: Equivalent Length

Nominal Pipe Size mm [in]	90° Long Sweep Elbow	45° Standard Elbow	Tee Flow on Branch	Tee Flow on Run	Male Adapter	Gate Valve
	Equivalent Pipe Length, m [ft]					
50 [2]	1.7 [5.7]	0.8 [2.6]	3.7 [12]	1.3 [4.3]	1.4 [4.5]	0.5 [1.5]
65 [2.5]	2.1 [6.9]	0.9 [3.1]	4.6 [15]	1.6 [5.1]	1.7 [5.5]	0.6 [2]
80 [3]	2.4 [7.9]	1.2 [4]	4.9 [16]	1.9 [6.2]	2 [6.5]	0.9 [3]
100 [4]	3.7 [12]	1.6 [5.1]	6.7 [22]	2.5 [8.3]	2.7 [9]	
150 [6]	5.5 [18]	2.4 [8]	10 [32.7]	3.8 [12.5]	4.3 [14]	

Table 5.5:
Equivalent length of common fittings used in rainwater conveyance piping. Adapted from www.engineeringtoolbox.com.

For each fitting in your system, add up the equivalent length to come up with your cumulative equivalent length of fittings (L_{eq}) in meters (feet).

Calculate Maximum Friction Loss

Next, you'll need the inner diameter of the pipe that you intend on using. If you are using schedule 40 PVC, you can pull it directly from Table 5.6. Otherwise, check with the manufacturer or supplier of your piping.

Now you have everything you need to estimate the maximum friction loss, in meters of water (feet of water) using the Hazen-Williams formula for full pipe flow.

Metric

$$\text{Max Friction Loss} = \frac{10.67 \times (L + L_{eq}) \times \left(\frac{Q}{C}\right)^{1.852}}{\dfrac{D^{4.871}}{1{,}000}}$$

US Customary

$$\text{Max Friction Loss} = \frac{10.46 \times (L + L_{eq}) \times \left(\frac{Q}{C}\right)^{1.852}}{D^{4.871}}$$

Where,
L: Length of pipe, m (ft)
L_{eq}: Cumulative equivalent length of fittings, m (ft)
Q: Max Flow rate, m/s (gpm)
C: Friction Coefficient
D: Pipe inner diameter, mm (in)

Table 5.6: Inner Diameter

Nominal Pipe Size, mm [in]	Inner Diameter, mm [in]
	Sch 40
50 [2]	52.5 [2.07]
65 [2.5]	62.7 [2.47]
80 [3]	77.9 [3.07]
100 [4]	102.3 [4.03]
150 [6]	154.1 [6.07]

Table 5.6:
Inner diameter for common piping used in rainwater conveyance systems. Adapted from www.engineeringtoolbox.com.

The Friction Coefficient is unit-less and is specific to the piping material you are using. For PVC pipe, use 140. If you are using a different material or different pipe thickness, you'll have to look up the associated friction coefficient and inner diameter, which is available quite readily on websites like engineeringtoolbox.com.

Head Pressure vs Maximum Friction Loss

You'll now take the maximum friction loss that you just calculated, in meters of water (feet of water), and compare that to your head pressure, in meters of water (feet of water).

It's simple at this point; for your wet conveyance system to work well throughout its entire operational range, you need the head pressure to be larger than the maximum friction loss.

Ensuring the Ability For Drainage

Make sure that you include some sort of drainage port or valve in the lowest point(s) of your wet conveyance system. You'll need this for periodic cleaning and maintenance and for ensuring water quality.

This can be difficult, or even impossible, if you choose to bury your wet conveyance piping, which is sometimes done to eliminate a tripping hazard, as shown in Figure 5.7.

This configuration will, over time, cause water-quality problems and may even plug off. We therefore recommend that you always plan for drainage in your wet conveyance piping. Consider, for instance, placing a small raised ramp above your wet conveyance piping to eliminate tripping hazards (rather than bury the piping). This way, you can continue to provide drainage via a small port accessible when you simply move the ramp.

If you look back to Figure 5.8 (underground wet conveyance system), you'll see that we were able to take advantage of landscape elevations to provide the ability for gravity drainage of the buried wet conveyance system.

Cold-Climate Snow Harvesting

While traveling through New Mexico, we visited a homestead where the owners were living entirely off rainwater. To increase the captured water in the winter, they had installed an electric heat mat under the roof. When there was snow, they would turn on the heat mat and use electricity to melt the snow and capture it as water. For their context — an acceptable cost for power and limited water availability — the owners felt that the gain in water was worth the electricity expended.

If this is something that makes sense for you, make sure that you carefully consider freeze and thaw protection for your pre-filtration equipment and conveyance piping.

Note, lastly, that wet conveyance systems should absolutely include a downspout filter and inlet/outlet screens to keep debris and insects out of the piping as much as possible.

Additional Cold-Climate Considerations

If you plan to use a wet conveyance system in a cold climate, you must either decide that it is a seasonal system and drain it before freezing temperatures every year, or you must provide frost protection. The same principles and options for frost protection of your wet conveyance piping apply as those suggested for tanks in Chapter 4, "Frost Protection and Layout."

Some added complexity arises in cold climates related to the interaction between your roof and your gutters. For instance, large snow loads on the roof can add substantial weight and even can rip your gutters right off your building when the snow slides off. A potential solution to this is installing specialized rails to prevent snow from sliding.

In addition, improperly considering the design of your attic ventilation and insulation can result in snow melt contributing to the formation of ice-dams and large icicles that form on roof edges and on your gutters. Gutters are not designed for this added weight, so it often results in damage.

Both of these issues are related more to good cold-climate construction practices than they are to cold-climate RWH systems, specifically. Nonetheless, if you are retrofitting an existing building, or building new and making choices about the design of your roof, it's wise to take the opportunity to do these things correctly, and save yourself cost and headache later down the road.

Chapter 6

Storage

Water Quality, Maintenance, and Your Tank

RESEARCH IS SHOWING that there is much more to your tank than what meets the eye, particularly when it comes to water quality.

Rainwater tanks actually support functional ecosystems comprised of complex communities of bacteria. The biofilm that forms on the internal walls of the tank and the sludge layer that forms over time on the bottom of the tank have beneficial implications for water quality (Spinks, 2007; Evans, et al., 2007).

Although some books and literature on maintaining rainwater tanks recommend that you de-sludge, clean, and sterilize your tank on a regular or annual basis, we don't recommend that you do this. To the contrary, in the sections below, we'll briefly discuss some of the biological and incidental processes going on in your tank and then make some recommendations on how you can best support these micro-ecologies to function and perform vital water-cleaning services for you.

Flocculation, Sludge, and Biofilms

Flocculation is the physical process in which particulate drops out of suspension and settles to the bottom of the tank. Over time, flocculated particulate accumulates at the bottom and forms the sludge layer, as shown in Figure 4.1, number 13.

The rate at which flocculation occurs depends on the particle size and the amount of disturbance in the water column of the tank. Think of it this way: the more the disturbance, the less flocculation there will be because more particulate will stay suspended in the water vs settling to the bottom.

Extremely high concentrations of heavy metals are found in the sludge layer of tanks and in the biofilm layer that forms on the inner tank walls, shown in Figure 4.1 as number 14 (Spinks, 2007). However, rather than see this as a bad thing, consider that this means that both the process of sedimentation and the biofilm remove heavy metals from the stored rainwater. The design impetus is therefore to encourage sedimentation and minimize re-suspension (disturbance) of the sludge and biofilm layers, as well as to draw water from above the sludge surface layer.

Design Directives to Support Sedimentation and Biofilm

Knowing that sludge and biofilm layers actually bio-accumulate harmful contaminants, especially heavy metals, you now have some impetus for design, particularly when it comes to setting up tank inlets and outlets. There are five design directives related to de-energizing the inlet water (and minimizing disturbance) that you should consider employing as much as possible. An example of a good inlet and outlet design configuration is shown in Figure 6.1 and described with corresponding numbers below:

i.   Keep a minimum water level in the tank above the inlet level. In the example provided this is done by fixing the intake a set height above tank inlet.

ii. De-energize water with the inlet piping configuration or inlet screens. In the example provided, flow is split into two and pointed upwards. A tank inlet filter/basket filter also turns concentrated flow into more of a rain shower.

iii. Place the inlet close to the center of the tank (as opposed to close to the tank wall) or, alternatively place the inlet and outlet at opposite sides of the tank.

iv. Larger tanks will be better at diffusing inlet energy than smaller ones.

v. Keep the intake above the sludge layer and avoid disturbing it. 100–200 mm (4–8 inches) is the recommended minimum height.

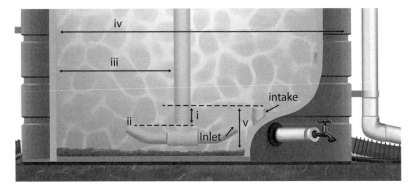

Fig. 6.1: *Five design directives to support flocculation, sedimentation, biofilm layers, and minimizing re-suspension of the sludge layer in your tank.*

Credit: Verge Permaculture/S. Andrei

Ideal vs Actual Tank Size

In Chapter 3, you established and optimized your total storage capacity based on your rainwater supply, your demand needs, and your minimum operating goals. If you are planning or able to use only one tank, your ideal tank volume is equal to this storage capacity.

If you have multiple tanks in multiple locations, you've already proportionally split up your total storage capacity based on the placement of your downspouts and the catchment area that each downspout receives. See Example 5.1, Splitting Up Tank Volumes Based on Downspouts, if you need to review how to do this.

This ideal tank volume (or these ideal tank volumes) are, of course, not necessarily representative of the actual size of tank that you'll end up with because you'll be rounding up to the nearest available size that the manufacturer offers. If building a tank, you'll be establishing your dimensions such that you end up with an actual volume equal to or slightly larger than your ideal volume.

Selecting Your Tank(s)

When it comes to selecting a tank from a manufacturer, your very best bet is to take your site plan, your elevation plan, your selected tank location(s), and your ideal tank volume(s). Now go and visit several local tank suppliers with this information, and leverage their expertise and recommendations, which are almost always provided free of charge. Despite the fact that suppliers are hoping to sell you a tank, we've found most to be incredibly knowledgeable about the products they carry, in addition to being honest and ethical if there's a better fit or a better solution (even if they don't carry it). Do a little research and find local expertise rather than trying to become an expert in all of the tank options available to you, as there are many!

If you are planning on building your own tank on your site, using a technique such as *ferrocement,* you'll most certainly be doing significantly more research. You'll also likely need construction experience and ambition as there are very few experienced ferrocement contractors in North America. We'll mention a few additional pros and cons about ferrocement tanks in the section below, but you'll want to see the book, *Water Storage, Tanks Cisterns, Aquifers, and Ponds* listed in the Resources section if you are seriously considering this as an option.

Note that in most of North America, regulators require tanks (and associated sealants and fittings) used in potable applications to meet NSF 61. However, research has shown that the water-quality impact of the tank material itself is small compared to the positive water-quality contributions of the tank micro-ecologies, such as the sludge and biofilm layers (Morrow, 2012). The exception to this is if lead was used in the construction of the tank, for instance lead solder or lead coatings on rivets.

What follows is a overview of the advantages and disadvantages of some of the most commonly used tanks and tank materials.

Pre-Cast Concrete Tanks

Pre-cast concrete tanks are those pre-built by a manufacturer and delivered in one piece to a site — by heavy equipment. They can also be cast in place, but we don't often see this for home-scale systems.

Concrete is made from gravel and limestone. As such, the material in of itself provides some benefits to water quality. For one, the walls will leach calcium carbonate into rainwater, and this mineral has benefits to health.

 Advantages:

- Long lasting (30 years +).
- Adds minerals, such as calcium carbonate, to the water.
- Impact on taste is marginal.
- Suitable for above-grade and one of better choices for buried applications.

 Disadvantages:

- Typically the most expensive option.
- Has a high embodied energy.
- Heavy and generally expensive and difficult to ship large distances.

Beware of Fly Ash When Purchasing Concrete

It should be noted that some concrete manufacturing uses fly ash from coal plants as a way of bulking up the concrete. According to the Wikipedia entry, fly ash can be made up a variety of substances which can include "arsenic, beryllium, boron, cadmium, chromium, hexavalent chromium, cobalt, lead, manganese, mercury, molybdenum, selenium, strontium, thallium and vanadium along with dioxins."

If you plan on using a concrete tank or making your own ferrocement tank, make sure that the concrete you source does not contain fly ash. Seek NSF 61-certified cement if you are designing for higher-quality water (such as drinking).

• Large equipment is needed (such as trucks and cranes) to put tank into its location. Not suitable for tight spaces or where access is an issue.

• May crack if not carefully designed and sited.

Ferrocement

Ferrocement tanks are custom-built onsite out of cement, rebar, and metal mesh, and they are usually built by the owners themselves. They are considered to provide the highest stored volume per dollar when storing large volumes of water in a tank, and hence the appeal for many homeowners.

Most of the ferrocement tanks we have seen exist in warmer regions of the world, including Australia, Mexico, California, and Washington State, and there is virtually no information available about using ferrocement in cold-climate applications.

 Advantages:

• Very long lasting (40 years +).

• Highest stored volume per dollar for large volumes of water.

• Adds minerals, such as calcium carbonate, to the water.

• Naturally neutralizes the pH of the stored water.

• Impact on taste is marginal.

• Fully customizable and tailor-made for site.

• Suitable for locations where space is constrained. No need for heavy equipment access (as compared to pre-cast concrete tanks).

• Suitable for above- or below-grade applications.

 Disadvantages:

• Very few skilled contractors.

• Most ferrocement tanks are owner-built and therefore require labor and ambition.

• Construction knowledge and skills are required, but opportunities to learn the technique are hard to come by.

• Not movable once built.

• Will require repairs from time to time.

Steel Tanks

There are a lot of different steel tanks out there. Depending on the size you are seeking, a steel tank may come manufactured in one piece or in modular units that are put together on your site. Tanks may also be lined with a membrane or have an inside coating. In North America, steel tanks can be used for drinking water applications if the liners or coatings meet NSF 61. In Australia, corrugated steel with a spray-on plastic food-grade polymer lining (Aquaplate) is quite common.

When researching steel tanks, consider looking beyond the conventional rainwater-supply places, and you might find a really good deal. We recently saw a local company that specializes in granaries (large steel containers used on farms for storing grain) offering a steel tank lined with a plastic membrane that would be ideal for rainwater harvesting.

We have also seen large metal rain tanks being fitted with rain gutters, allowing for combined storage and collection of rainwater.

One very important point with steel tanks is to not connect metal piping to a metal tank. This will create a corrosion circuit and destroy your tank and/or plumbing. A better practice is to use poly plumbing for at least the first two meters after the tank. As always, be sure that the poly piping and components are potable-rated.

Lastly, be sure to get the manufacturer to confirm that there was no lead used in the construction of the tank.

 Advantages:

• Long lasting (20–30 years +).

• Reasonably cost effective, often a good choice for mid-sized above-grade tanks.

- Available in a decent range of storage volumes.
- Impact on taste is marginal.

 Disadvantages:

- Not usually acceptable below grade.

Stainless Steel

Stainless steel is considered highly inert; it is used in surgical and food processing applications due to its safe nature. While there are not a lot of stainless steel rainwater tank suppliers globally, it is a material that we saw used in Australia and is considered by some to be one of the premium rainwater tank materials.

However, it is crucially important to be aware of issues that may arise with corrosion circuits if metal piping is connected to any steel tank. As such, the same recommendations regarding the use of poly plumbing/piping given in the previous section apply. Also, be sure to get the manufacturer to confirm that there was no lead used in its construction.

 Advantages:

- Very long lasting (100 years +).
- Resists corrosion internally and externally.
- Easy to clean.
- Virtually no impact on taste.

 Disadvantages:

- Expensive.
- Bulky and/or difficult to ship.
- Typically only used above grade.

Fiberglass

Fiberglass tanks are lightweight, strong, and corrosion resistant. Overall, fiberglass has decent properties for rainwater storage for both above- and below-grade applications. It's recommended that the interior resins and/or coatings used meet NSF 61.

We have heard reports of increased contamination from this type of tank, but at the time of publication didn't have access to research substantiating this. As always, do your own research and ask good questions of the supplier/manufacturer. We'll update our website as we learn more as well.

 Advantages:

- Lightweight, easy to transport, easy to move.
- Available in many sizes and colors.
- Can be used above or below grade.
- Fittings are integral to the tank and therefore don't leak.
- Easy to repair.

 Disadvantages:

- For small sizes (1,000 gallons), fiberglass tanks are not usually cost effective.

Plastic/Poly Tanks

There are quite a few different types of plastic tanks specially for small to medium tank sizes. These are often the best choice for small to medium, low-cost water storage. You'll want to go straight to your supplier to know what is available and the pros and cons of each model and, if need be, confirm that the model you select meets potable water specifications.

Grandma's Cistern

Michelle's mother, who lived in northern Saskatchewan, grew up in a home that had an integrated rainwater cistern in the basement. The cistern was basically a closed-off room: one of the vertical walls was the concrete wall of the basement foundation and the other three walls were made from brick, forming about 16,000 liters (4,227 gal) of water storage. It's a good example of how to design year-round water storage in a cold climate by integrating the cistern into the design of the basement. While this particular cistern was sealed off using a lining of tar, present-day replications of this design should stick to coatings or liners certified to NSF 61.

Manufacturers do make plastic tanks that are acceptable for below-grade applications. However, there may be special burial procedures, materials, and maximum depths.

As always, check the materials and processes used to construct the tank.

 Advantages:

- Medium length life span (15 years +)
- Relatively low cost.
- Come in many shapes and sizes, large and small.
- Lightweight to transport.
- Readily available off-the-shelf in many places.
- Higher negative impact on taste than concrete and steel tanks.

 Disadvantages:

- Tough to repair.
- May degrade in the sun. Best placed in a shady location.

Bladder Style

There are some soft-membrane tanks available on the market that are sometimes referred to as *bladder tanks*. These flexible plastic-based containers are low-profile and lie flat, quite like a waterbed mattress.

In addition to being relatively inexpensive, the appeal is that these tanks can be placed in an otherwise unused and difficult-to-access space, such as a heated crawlspace under a home, or under a deck. While leaks, ruptures, and drainage are important design considerations for the placement of all types of tanks (see "Thinking About Spatial Layout" in Chapter 4), the higher likelihood of placing bladder-style tanks in high-risk locations requires increased design emphasis on a tank-rupture scenario. Combine a high-risk location with the questionable life

expectancy of the seams and fittings on bladder-style tanks, and we don't recommend using this type of tank without the use of secondary containment and the provision of somewhere safe for the water to go.

To give you an example, if you wanted to place a bladder-style tank in your crawlspace, you should consider also adding an extra pond liner and drain or emergency drainage design such that if (or when?) the bladder tank fails, the large volume of water that is released is contained and is directed appropriately to prevent damage to your building or cause mold problems.

By the time you account for secondary containment, it may not be as cost-effective as originally thought. However, you may still consider using one — especially if you have very limited outdoor space or if you required freeze protection and the cost to bury a tank onsite is very high.

 Advantages:

- Low cost for the bladder tank itself.
- Ability to fit into tight spaces or use otherwise unused space.
- Very easy to transport and ship.
- Higher negative impact on taste than concrete and steel tanks.

 Disadvantages:

- Check life span with manufacturer and be sure to consider this in design and placement. May require secondary containment.
- May degrade in the sun. Best placed in a shady location.
- Not suitable for below-grade applications.
- Lack of breathability may result in lower water quality.

Tank Component Design

Here we'll go through the design details of all of the major components in your rainwater tank (with the exception of the inlet, which was covered in the section "Design Directives to Support Sedimentation and Biofilm.")

All reference numbers in the section below refer to the numbers indicated on Figure 4.1.

Intake and Outlet

The intake and the outlet port are shown as numbers 15 and 16, respectively.

The outlet port is a bulkhead fitting, or sometimes a welded fitting provided by the tank manufacturer. This fitting allows for a connection to pass through the tank wall without allowing any leaks, usually by using a gasket to form a liquid-tight seal on the tank wall. It is usually placed very near to the bottom of the tank.

On the inside of the tank, the intake (and associated piping) is usually connected to the outlet port in one of two ways: floating or static. A floating intake (number 15) uses a flexible pipe connected to a buoy which keeps the intake suspended near the middle of the column of water. As the tank level drops, the intake drops with the water column.

A static intake is usually an upturned 90° PVC elbow connected to 100–200 mm (4–8 in) of rigid PVC pipe. With a static intake, you are always drawing water from near the bottom of the tank, but you place the inlet of the PVC pipe high enough to not disturb the sludge layer (see Figure 6.1). It's not recommended that you glue or weld the elbow to the outlet port and/or to the PVC pipe. You'll want to maintain the ability to rotate this elbow to allow for full drainage and/or disassembly.

On the outside wall of the tank, you'll typically connect your outlet port to piping that is configured dependent on your end-use, which may or may not include a pump. This piping may also be connected to another tank, which is discussed in the section called "Connecting Multiple Tanks," later in this chapter.

If you intend on connecting your tank to a pump, it is recommended that you fit an inlet filter upstream of the pump to protect it. You can do this by using a screen on the intake inlet or by installing a Y-strainer filter (number 26) before the pump. An Y-strainer is easier to service, and hence, our preferred choice for a pre-pump filter.

Regardless of your piping configuration, it's a very good idea to place a valve immediately downstream of the tank outlet port (called an *isolation valve,* which is shown as number 24). If the pump, downstream tanks, or other fixtures ever need to be serviced or replaced, this valve allows for maintenance without the need to drain the tank.

Overflow

Our preferred overflow configuration on the inside of the tank wall is to have an upturned 90° PVC elbow and a small upward-pointing section of rigid PVC pipe (number 17). The length of this pipe sets the maximum height of water in the tank. The overflow port, the piping inside, and tank (and all downstream overflow piping) should have at least the same diameter as the inlet (number 11), ensuring that overflow does not cause a flow bottleneck in the system. A skimmer is also shown and is simply a 45° angle cut into the overflow inlet. It's a good design practice because it allows dust and floating particulate to more easily enter the overflow inlet, and so more easily exit your tank.

The overflow port (number 18) is a fitting, often a bulkhead fitting, the purpose of which is to allow for a connection to pass through the tank wall without allowing any leaks. It is usually placed on a vertical wall near the top of the tank.

On the outside of the tank wall, you'll connect the overflow port to piping designed to carry water to its final landscape location (you may also use this to connect to another tank — more on that later). Overflow piping is shown as number 19. Make sure that this piping maintains positive drainage by keeping at least a 1%–2% downward slope. Also, prevent vermin, insects, and larva from getting into the tank by including an outlet with a screen, or equivalent.

Of primary concern is that overflow water is directed such that it can ultimately gravity flow away from the tank foundation or any other infrastructure that might be damaged by water. However, you'll miss out on a huge opportunity if you don't take this one step further and actually consider how to put this overflow water to productive use in your landscape through the use of water-harvesting earthworks, shown as number 20.

Rainwater Harvesting for Drylands and Beyond, Vol II and *Permaculture: A Designers' Manual* are two excellent resources, should you want to explore water-harvesting earthworks in more detail. Both are listed in the Resources section of this book.

Rainwater Overflows on Our Urban Property

On our home we have installed a three-season RWH system for outdoor irrigation complete with diversion valve, first flush diverter, and 5,000 liters (1,321 gal) of total water storage between three tanks in three different locations.

The overflow water from the tanks is redirected to a landscape water-harvesting feature that we call *swale trails*. This is a variation on the typical permaculture swale and is basically perforated tubing (also called *Big O*) that we dug and laid down, on contour, through our front-yard food forest and backyard garden. We then covered this perforated tubing with mulch to form pathways through the garden and food forest. Because of this, there is virtually no stormwater runoff from our property, and all water that lands on our property is put to productive use. For more information on how we installed these urban swales, head to www.vergepermaculture.ca and search "swales," or view the mini-documentary about our home and garden at www.vergepermaculture.ca/meet-our-team/.

Fig. 6.2: *The authors' home in 2009, before starting on their property renovation.*

Fig. 6.3: *Water from one garage downspout is directly connected to perforated tubing that runs through one section of the garden.* CREDIT: VERGE PERMACULTURE/G. YOUNG

Fig. 6.4: *The food forest in the front yard of the authors' home is entirely irrigated by overflow through their RWH tanks and the neighbor's downspout.*

CREDIT: VERGE PERMACULTURE/G. YOUNG

Fig. 6.5: *A close-up of the neighbor's downspout being redirected to the authors' food forest.*

CREDIT: VERGE PERMACULTURE/G. YOUNG

Fig. 6.6: *Michelle's fully rainwater-irrigated vegetable garden.*

CREDIT: VERGE PERMACULTURE/G. YOUNG

Fig. 6.7: *The authors' children enjoying the bounty and beauty of the garden, which is passively irrigated with rainwater.* PHOTO CREDIT: VERGE PERMACULTURE/G. YOUNG

Level Indicator

A level indicator is an important feedback mechanism that provides the user with information about storage volume as well as usage patterns. It is shown as number 21.

There are several tank level indicators on the market. Some operate using a float, others ultrasonic measurement, and some use a pressure measurement. Selecting a level indicator is typically a question of service requirements, tank dimensions/geometry, what's readily available, and cost. Be sure to review the manufacturers' specifications to ensure that the indicator you select will work for your scenario.

Note that many level indicators need to be installed *during* construction and cannot be installed easily post-tank commissioning.

Manway and Air Vents

A manway (number 22) is a hatch that is used for access and maintenance.

The installation of components inside the tank, including the quiet inlet, the intake piping, overflow piping, and level indicator may require someone to enter the tank; therefore, the hatch should be large enough to allow for safe entry and egress. When you are designing

and selecting the interior tank components, make sure to keep this installation requirement in mind. For instance, consider how the quiet inlet piping (number 11) must be lowered in through the manway, then connected to the inlet port (number 12). The inlet port is very likely a threaded bulkhead fitting and therefore whoever is installing it will need to be able to hold up the quiet inlet piping, line it up with the fitting, then thread the entire assembly into place — not an easy maneuver in a tight space (we learned this through direct experience).

An air vent (number 23) is also a very important component. It ensures that air can enter and leave the tank as the water level in the tank changes. Most off-the-shelf rainwater tanks have built-in air vents integrated into the manway of the tank. The vent should be screened to ensure that no insects or other critters can enter the tank.

Foundations and Securing Your Tanks

The design and installation of your tank foundation is a crucial step for any RWH system. As the title of this book suggests, however, here we'll only provide you with some of the *essential* considerations for foundations. It's a much bigger topic than we can cover here.

Don't underestimate the weight of water! Consider the 1,000 liter (265 gallons) tote shown in Figure 6.8. We see these frequently in urban irrigation applications, as they are relatively cheap and easy to find used. When full of water, this tote will weight 1 metric tonne (2,200 lbs). And this is considered very small compared to the rain tanks often needed for primary supply and off-grid supply scenarios.

An improperly constructed foundation can result in all sorts of problems, including the tank settling below ground level, water pooling, uneven settling causing your tank to lean,

Tanks Are Confined Spaces

It should be noted that rainwater tanks, above- or below-grade, are considered confined spaces and can very dangerous because of lack of oxygen, poor air quality, physical hazards, and more. Beyond entry that may be required for installation, be sure to consider what you can do, from a design perspective, to minimize the need for entry in the future.

If and/or when the tank does need to be entered, be sure to review the best practices and safety procedures for enclosed or confined entry and take appropriate precautions seriously. You may even consider hiring a professional installer.

damage to your tank (including cracked walls and floors), and damage to associated piping connections.

Tanks that have a high height-to-footing ratio (often called *slimline*) are good choices for tight spaces because they are slim and fit well alongside buildings. However, their center of gravity is high; therefore, settling in the foundation can result in the tank tipping over, which can cause significant damage, especially if it tips toward infrastructure, such as a wall.

Foundations can be built a number of ways depending on your goals, budget, needs, and site constraints. Above-grade foundations include gravel pads, concrete slabs, and raised decks on steel or concrete piles. The most common type of foundation we see used by do-it-yourself owner-builders are gravel pad foundations, which are relatively simple to build and low cost.

If you have no experience designing or installing foundations, you'll absolutely want to work with a professional installer. Note also that if you purchase an off-the-shelf tank, the manufacturer will have installation requirements related to the foundation that are specific to the tank dimensions and the tank materials (steel vs plastic, for instance). These requirements are not

Fig. 6.8: *A 1,000 liter (265 gal) tote will weigh 1 metric ton (2,200 lbs) when full of water. The foundation for this tank is a gravel base topped with an industrial-grade plastic pallet (salvaged from a construction site). Note that if you decide to seek out a used tote like this one, ensure that the previous contents were nontoxic.*

Fig. 6.9: *Multiple slimline tanks adjacent to a building on a gravel pad foundation.*

only critical to the proper functioning of your system, but they also ensure that you don't void any warranty.

Here are some very basic principles for building a solid gravel base (number 28):

- Excavate and dig out all organics (all topsoil) until you reach firm, consolidated subsoils.
- The footprint of the excavated area should be slightly larger than the footprint of the tank, typically by at least 15 cm (6 in).
- Make the excavation as level as possible and remove any large rocks.
- Consider laying down a geo-textile. It will prevent the migration of the fill you are about to put into the hole, and may also allow for a shallower excavation depth (in the case of very deep top soils).
- Talk to your local gravel supply company about the products (also called "fill") that they carry that are *compactable*. The idea here is that you need to place something in the hole, and

compact it such that it will not settle further (or settle very little) over time.

- Refill the excavation with 10–15 cm (4–6 in) layers of fill at a time. Use a compaction device such as vibratory plate compactor to ensure that the layer is fully compacted before adding another layer. These machines save a ton of time over hand-tamping devices and are probably available for rent at your nearest construction equipment rental shop.
- Have your gravel pad protrude above the top surface of the ground and consider your drainage planes. You want to set a slight slope for drainage away from other infrastructure, such as buildings.
- The final layer of the foundation should not puncture or compromise the tank. Smooth gravel, geo-textile, pea-gravel, and even used synthetic carpet are good options for this final layer.

There are other ways to prepare foundations, none of which we will cover here. *Water Storage*

Fig. 6.10:
Compacting the fill layer for an above-grade tank using a vibratory plate compactor.

(Ludwig, 2013) is an excellent resource if you are planning on constructing your own tank. It is included in the Resources section of this book.

If you are planning on raising your tank up onto a tank stand, all of the same principles apply regarding construction and the need for stability of the base. This is not a small undertaking, so be sure to hire an engineer or good contractor if you are not experienced yourself.

Lastly, when empty, plastic tanks in particular can be quite lightweight. If you live in a windy region, you may want to consider use guy-wire or other means to anchor your tank to the ground to prevent it from being moved by wind.

Supplying Make-Up Water

You may want to consider having the ability to supplement your rainwater with an alternative water-supply source. This *make-up water* can come from a pressurized water source, like municipal water or a groundwater well, or it could come via a delivery truck.

There are a few different possible configurations for tying in your make-up water to your RWH system. The water can be supplied via a bypass connected directly to the distribution system (downstream of your rainwater pump), or, alternatively, by directly adding water to your tank (upstream of your rainwater pump). If you are using pressurized water as your make-up water, such as municipal supply, one advantage of connecting your bypass downstream of the rainwater pump is that you don't waste all that pressurized energy. Consider the alternative where you take high-pressure municipal water and de-energize it in your tank, only to need to re-energize it again through your rainwater pump.

Make-up water systems can also be controlled manually (e.g. the user manually turns on or connects a hose), or they can be automatic, where additional float switches or other sensing devices are used to control switches or three-way valves.

If you are providing make-up water from the municipal system or other community system, nearly all jurisdictions and regulators will require that your design include some sort of backflow protection, meaning that you must ensure there is no way for your rainwater to flow backward and potentially contaminate the make-up system.

There are manufacturers who have designed elegant all-in-one units that use reverse pressure principles to provide both automatic switching and backflow prevention. These are ideal when make-up water is tied-in downstream of the rainwater pump. If you use this option, make sure that the valve is set to preferentially use rainwater when it is available, not the other way around.

When the make-up water is tied-in upstream of the rainwater pump, a simpler combination using a specialized top-up valve and air gap can be used. The specialized valve provides automatic filling of the tank with make-up water, and the air gap prevents the possibility of backward contamination. This combination is illustrated in Figure 6.11.

Note that there are actually two air gaps required in the Figure 6.11 configuration to completely eliminate the risk of backflow contamination and for good design. The first air gap "A" is the distance between the overflow and the top-up valve. The second air gap "B" is important in the event of plugging off or backup in the overflow piping. Air gap "B" is also very important to ensuring that the top-up valve is working properly, for instance visually ensuring the valve is not stuck in the open position, causing ongoing overflow and a complete waste of water.

The rules around what is permissible and required for backflow prevention may vary from jurisdiction to jurisdiction; therefore, be sure to consult a plumbing professional or your local plumbing code before finalizing your design.

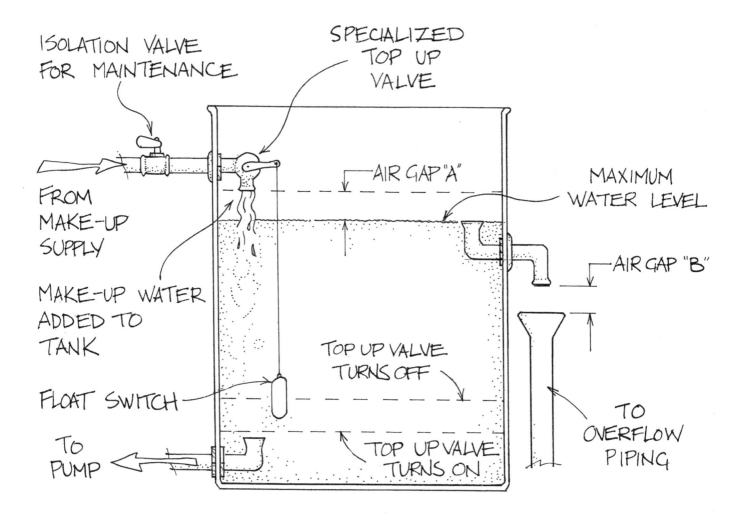

ISOLATION VALVE
FOR MAINTENANCE

SPECIALIZED
TOP UP
VALVE

AIR GAP "A"

MAXIMUM
WATER LEVEL

FROM
MAKE-UP
SUPPLY

AIR GAP "B"

MAKE-UP WATER
ADDED TO
TANK

TOP UP VALVE
TURNS OFF

FLOAT SWITCH

TO
OVERFLOW
PIPING

TO
PUMP

TOP UP VALVE
TURNS ON

Fig. 6.11: *A specialized top-up valve and air gap combination allow for automatic filling of the tank with make-up water while preventing the possibility of rainwater contaminating the make-up water supply.*

Connecting Multiple Tanks

When connecting multiple tanks, it is important to design the connection details in such a way that the tanks can move independently. To avoid any interconnected piping from shearing in the event tanks settle or shift, use a flex joint like the one pictured in Figure 6.12.

If you intend on connecting multiple tanks there are two ways to accomplish this: in series or in parallel.

Connecting Tanks in Parallel

Tanks that are set up in parallel are connected via the piping downstream of the isolation valve (number 24).

In this configuration, the water level in all tanks will be the same and the last tank in the line is the only one to include an overflow outlet. See Figure 6.13.

Note that because of this shared water level and shared overflow system, setting parallel tanks up at different elevations will either cause overflow issues or will reduce the capacity of the tank at the highest elevation.

Connecting Tanks in Series

Tanks that are set up in series are connected via the piping downstream of the overflow port (number 18).

This configuration works on a cascading principle which means that the first tank will fill then overflow or "cascade" into the second and

Fig. 6.12:
Flex joint between connected tanks.
CREDIT:
VERGE PERMACULTURE

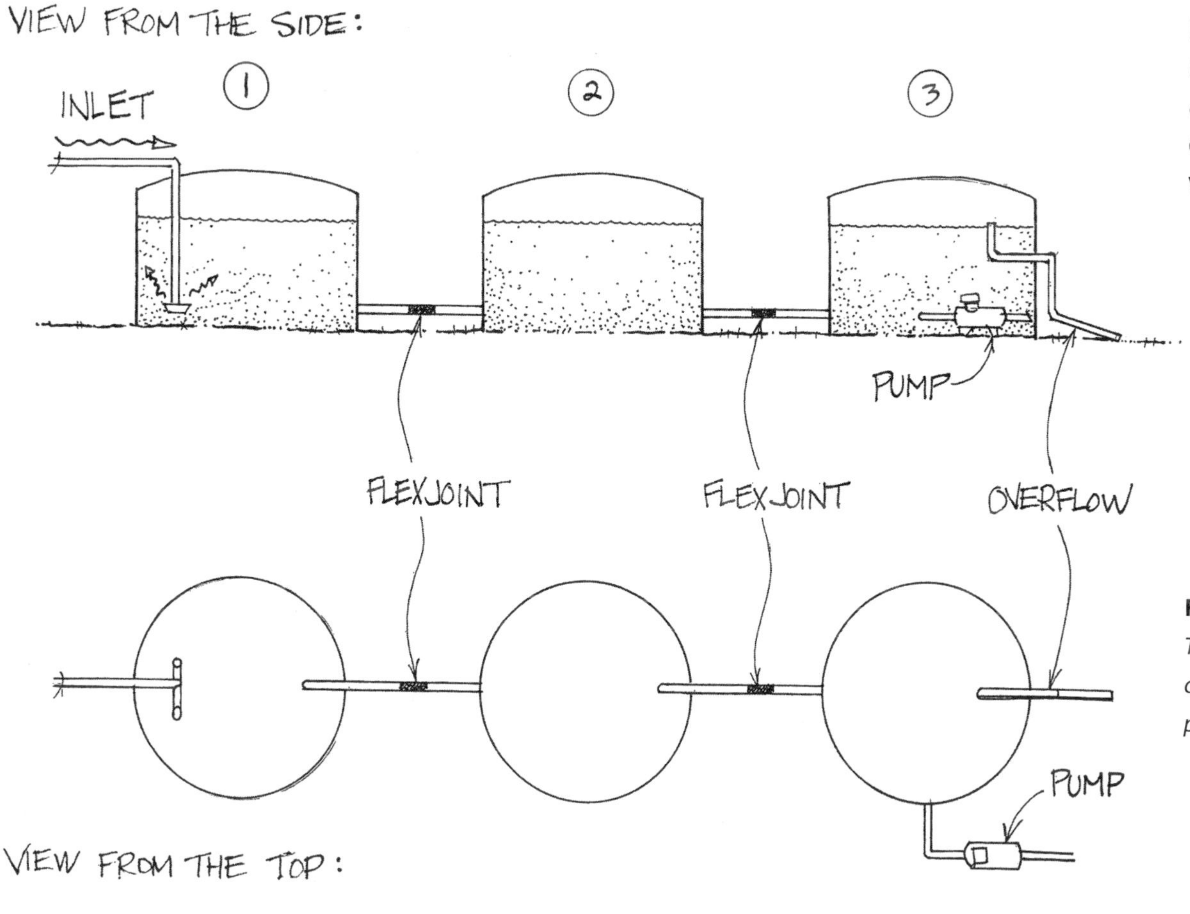

VIEW FROM THE SIDE:

INLET ① ② ③

FLEXJOINT   FLEXJOINT   OVERFLOW

PUMP

Fig. 6.13:
Three tanks connected in parallel.

VIEW FROM THE TOP:

PUMP

so on and so forth. Cascading systems are great for creating long settling times, meaning higher sedimentation and a higher water quality at the final outlet.

With in-series configurations, an operational challenge arises when stored water volumes are low. Refer to Figure 6.14 and note that if you are relying on the overflow connections to connect your tanks, you'll have no way to access the water in tank 1 or tank 2 once tank 3 is empty. Therefore, this configuration requires the use of secondary bypass pipe and associated valves, which would be configured precisely as you would if you were connecting tanks in parallel. As such, there is a benefit to this configuration, particularly when water quality is important; however, you double the amount of inter-tank piping connections that are required because you essentially need to set up both in series and parallel piping because of operational constraints.

In-series tank configurations can either be set up at the same elevation, or tanks farther along the cascade can be set at lower elevations. If you do it the other way around (tanks farther along the cascade are higher than previous tanks), then you'll eliminate your ability to gravity flow water through your bypass piping toward the outlet.

Note that Figure 6.14 shows the overflow piping in tank 1 and tank 2 pointing downward. The intent here is that the tank 1 and 2 overflow water is drawn from the cleanest part of the water column (closer to the middle) allowing for the highest-quality water to end up in tank 3. However, if you use downward pointing overflow piping, be sure to set up the piping with an anti-siphon vent.

Protecting Stored Water from Freezing

If you need to provide frost protection for your tank, go back to Chapter 4 and review the section "Frost Protection and Layout."

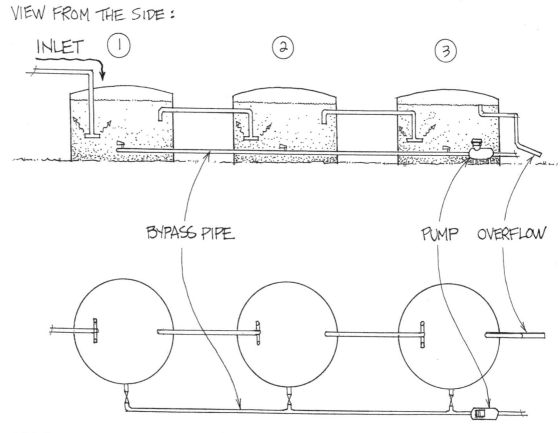

Fig. 6.14: *Three tanks connected in series with bypass piping.*

Below-Grade Tank Design

Below-grade installation adds considerable complexity to your design. Here we will only provide some general considerations to help you decide if you might explore this direction further with your local professionals such as civil engineers, excavating professionals, and/or tank suppliers and manufacturers.

Site-Specific Considerations
Building Codes

There will likely be additional local building codes, permits, and regulations pertaining to the both the design and the safe installation of underground infrastructure.

Water Table

The water table is a major consideration because of groundwater-generated buoyancy forces. We've actually witnessed very large tanks completely "pushed" out of the ground post-installation due to groundwater. If you don't know, or understand, the groundwater levels on your site (including the potential for seasonal variation), you should consult a professional. You may be required to design and/or add ballast (i.e. weight) to your tank to counteract these buoyancy forces and keep the tank underground.

Soil Composition

The composition of your soil could add cost and complexity to the installation of an underground tank. A test hole should be dug by the installation contractor to identify any potential installation issues related to the soils.

Installation Considerations

If you want a below-grade tank, it must be rated for that application. Manufacturers will also have installation requirements for maximum burial depth, backfill, and the design of the foundation.

Don't underestimate the cost of hiring an excavator, and potentially a crane, and the space required for the equipment to perform the installation. Early on in your planning you'll want to be talking to an excavation company about these considerations. Something else to consider is that getting rid of clean fill can be expensive. Use the fill on site, if possible.

You'll also likely have associated underground piping. As with above-grade tanks, a proper foundation is critical to ensuring that the tank doesn't settle because settling can easily break any piping connections. Another common construction technique used for below-grade piping is to use non-compatible, free-flowing gravel that will move around the pipe if the soil below the piping should settle. Lastly, bury piping with a small lift of appropriate backfill material, then place a piece of dimensional pressure-treated lumber and/or a strip of burial caution tape before completing the remaining backfill. The idea here is that if someone digs in the future, they may see the caution tape (or hit the wood) and stop digging before hitting the actual piping.

Also, when installing any underground infrastructure (including pipes and irrigation lines), we highly recommend that you properly record the location on a final, "as-built" site plan. It is also good practice to install trace wire on non-metallic underground piping to make it easier for future line locating.

We strongly emphasis good installation practices for underground infrastructure, particularly because it is a pervasive legacy issue that we run into on client properties. The costs and safety risks can be high.

Essential Components of a Below-Grade System

A Protruding Manway

It is important for the below-grade tank to have a manway that protrudes above the ground elevation (number 1 in Figure 6.15). Check your regulations. There may be a minimum protrusion height. The manway must be properly sealed and covered to prevent ground water from entering into the tank.

Beyond that, the same considerations for manways apply that have already been described.

Overflow

The same overflow fundamentals apply to underground tanks as they do to above-grade tanks. First off, water overflows must be directed away from the tank. Secondly, the tank overflow port must be at a high enough elevation such that water can gravity flow to the proper landscape location. Because the tank is buried and thus deeper than the surface elevation, this is typically managed by sending the overflow (or overflow piping) up into the tank manway.

Depending on site elevations, there are a few ways to do this. One such way is shown on Figure 6.15 as number 2. The overflow piping

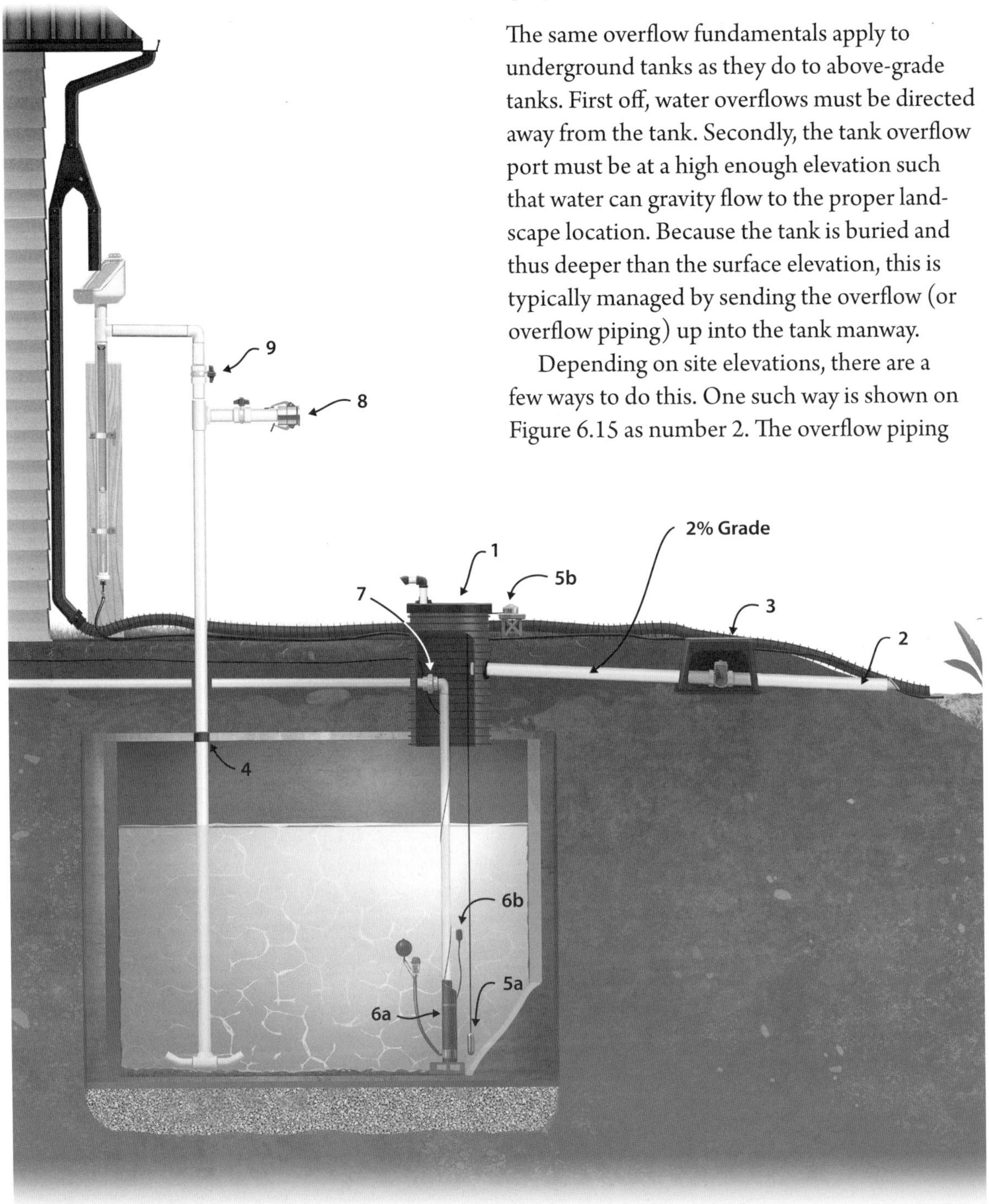

Fig. 6.15:

An example of a below-grade RWH system.

CREDIT:

VERGE PERMACULTURE/

S. ANDREI

can stay below grade because the final landscape location is at a low enough elevation.

Alternatively, the overflow piping may have to protrude up and outward from the manway, fully above grade. See Figure 6.16.

It's important to ensure that surface water cannot back flow up the overflow pipe. Ensure proper drainage with a 1–2% minimum grade; have enough clearance at the outfall and/or by installing a check valve, with access box (number 3 on Figure 6.15).

We've seen tanks overflow designed to be reliant on a separate pump, but do not think that this is good design practice.

Pipe Connections

Below-grade tanks of all types must have specialized water-tight fittings for any connection that passes through the tank wall. The placement of one such fitting is shown as number 4.

Not only do these fittings prevent groundwater from seeping into your tank, they may prevent other contaminants from entering your tank, for instance, effluent from a nearby septic system.

Level Sensing Devices

There are several different options for level sensing. One is a *pressure transducer* (number 5a) and *transmitter* (number 5b) that correlates head pressure to water level. Many other options exist, including ultrasonic measurement, float mast, and the low-tech dipstick.

Accessing the Water

Getting access to water in a below-grade tank is more complex than above-grade tanks because the water is typically below where you want to use it. For this reason, you'll almost always require a lifting device like a hand pump or an electric pump.

The pump can be a submersible model, or it can be an external pump and not be submerged

in the tank water itself. The submersible pump shown (number 6a) comes equipped with a floating and screened intake pipe.

Always consider how you intend providing maintenance to your pump and associated intake screens. In Figure 6.15, a union fitting is installed in the manway (number 7) allowing for piping to be disassembled and for the pump (and associated filter) to be pulled out for servicing. Another common technique is to use a fitting called a pitless adapter.

Number 6b is the run-dry protection on the pump, which is discussed in the next chapter.

Supplemental Fill Option

In the below-grade tank example provided, number 8 is a manual supplemental fill port with a quick cam lock fitting for adding trucked water to the storage tank in times of rain shortages. Number 9 is a valve that is usually in the open position, but closed when water is being pumped from the water-hauling truck.

Fig. 6.16: *Example of overflow piping directed from a protruding manway. This tank overflow is supplying water to a swale on contour that is planted with a raspberry hedge.* Credit: Verge Permaculture

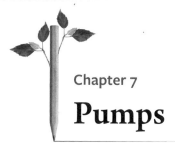

Chapter 7

Pumps

THE WORLD OF PUMPS IS VAST. A whole book could be written on piping, pressure loss, and correct pump sizing. But, as with the selection of tanks, we don't think that you need to become an expert in pump sizing and selection in order to design your own RWH system.

In this short chapter, we are going to focus on giving you some basic knowledge of pumps. We recommend that you use this simply as a starting place to help you narrow down your options before going to talk to a pump specialist/supplier. Pump suppliers are incredibly knowledgeable and will most certainly steer you in the right direction. Even though we are experienced, we often rely on suppliers to help us with sizing and selection.

The three main things that you need to make a decision on are:

i.   Is your pump going be external (outside of the tank) or will it be submersible (submerged in the water tank itself)?

ii. What control strategy will you use (manual, automatic, single speed, variable speed)?

iii. What type of pump will you use (centrifugal, multi-stage, jet, etc.)?

There are also some standard installation practices that you'll want to follow, in addition to any manufacturer-specific requirements and recommendations.

You'll want to collect the following information before talking to your pump supplier:

• What you are trying to do? Are you supplying irrigation, toilet flushing, house service, etc.?

• How much flow do you need? If you don't know this, they should be able to help you.

• What diameter pipe are you planning on using for your distribution piping? If you don't know, they can likely make some recommendations.

• What is the material of your distribution piping? Some examples include copper, PEX, and PVC.

• What are the pressure requirements of the end-use, for instance required domestic supply or irrigation supply pressure?

• What is the power supply? Is it AC or DC? If using AC, what is the supply voltage?

• Are you on-grid or off-grid? If off-grid, is there a max inrush current that your inverter can handle?

In addition to these answers, your site plan and your elevation plan view drawing are particularly useful tools that you'll want to share with the supplier.

External vs Submersible Pumps

Pumps are designed to either be external to the liquid being pumped or to be submerged in the fluid they are pumping. External pumps are generally easier to access and therefore service.

For below-grade tanks, the decision to go external versus submersible will likely made by you in consultation with your pump supplier. The primary motivating reason to use an external pump, installed at or near the surface, is again for ease of maintenance and servicing. However, the greater the elevation distance between the pump and the surface of the water in your underground tank, the greater the risk of *cavitation* (vapor bubbles exploding within the pump inlet) which can be extremely damaging to your pump. Given that you've already prepared your elevation plan view drawing in Chapter 4, you'll already have the information you need to help your supplier help you to make this decision.

Control Strategy, Single Speed and Variable Speed

Pumps require a control strategy. You need to decide how your pump will turn on and, in the event of an automated strategy, what signal (pressure, flow, moisture, or time) drives the pump to energize.

It's also worth understanding the difference between single speed pumps and variable speed pumps. The first has only two operating modes: on or off. The latter can start slowly and ramp up capacity over time (often called *soft start*).

A manual switch strategy is the simplest of all. Basically, when you want the water you turn it on, when you are done, you turn it off.

A timer or moisture control strategy would typically be used with an irrigation system. Timers will turn the pump on using a predefined schedule, or a soil moisture measurement can be used to turn the pump on/off.

The most commonly used strategy is a pressure-driven single-speed pump. This configuration requires a pressure transducer and a bladder pressure tank located downstream of the pump. When the pressure drops, the pump turns on. It turns off again when a set pressure is regained. The bladder pressure tank helps to maintain water pressure, minimizes pump cycling, and it protects against water hammer (a pressure wave that can cause equipment damage that is caused by a quick change of water velocity).

More and more, variable speed pumps with soft starts are being deployed in residential applications because the cost has come down significantly in recent years. These pumps use flow, pressure, or a combination of both to dictate the speed at which the pump operates. Although more expensive than single-speed versions, they are also substantially more energy efficient and are much better in off-grid power systems (a single-speed pump requires large in-rush currents, which can overwhelm inverters). Many models don't require a bladder pressure tank.

Types of Pumps

There are three types of pumps that are most often used in RWH systems.

Single Stage Centrifugal (SSC) pumps push water by spinning an impeller at high RPM through the use of centrifugal force. They are generally inexpensive, versatile, and reliable. SSC pumps come in external and submersible versions and are available with single- or variable-speed controls. SSC pumps are generally used for higher flow, lower pressure applications like irrigation.

Multistage Centrifugal (MC) pumps are used for high-flow, higher-pressure situations; they are typically used in submersible applications. The pump shown in Figure 6.15 is an MC pump.

Jet pumps are centrifugal pumps that take a slip-stream of water from the pump outlet and re-inject this slipstream back into the inlet side of the pump, though a constricted section of pipe (called a *venturi*). This constriction creates a vacuum (suction) on the inlet side of the pump. The pump shown in Figure 4.1 is a jet pump.

All pumps will have operating ranges (and limitations) based on inlet suction, horsepower, flow rate, and delivery pressure.

If your RWH system is for drinking or high-quality end-uses, look for a pump certified to potable standards (NSF 61 for US and Canada).

Lastly, note that most water uses inside the house (and even irrigation) require relatively low flow rates — so don't oversize your pump.

Installation Recommendations

When installing a pump, it is worth observing a few best practices to ensure that your pump operates correctly and that future diagnostics and servicing is easy.

1) Keep suction piping as short as possible (5–10 times the diameter of the pipe). Minimizing pressure loss in the suction side of the pump is important for proper pump operation.

2) Pipe diameter on the suction side should be equal to or one size larger than the pump inlet to minimize friction loss and velocity.

3) Eliminate elbows (90° or 45°) or tees mounted on or close to the inlet nozzle of the pump, as these create turbulence and pressure losses. Keep the inlet piping as straight as possible.

4) Eliminate the potential for air entrapment in suction piping. Make sure that pipes are properly sealed, glued, or otherwise connected.

5) Don't run your pump if your tank is near empty or empty. *Run dry protection* is a feature usually available from your supplier and is highly recommended.

6) Ensure the piping arrangement does not cause strain on the pump casing. Pipes should be designed to self support. Do not use the pump as a pipe support.

7) Pressure gauges are great diagnostic tools and will give you insights into how the pump is operating in normal conditions as well as insight when it is not operating optimally.

8) Setting pumps up with isolation valves and unions can make servicing the pump easier in the future.

9) Placing a drain point into the low points of the system will ensure that when the pump is isolated, pressurized fluid can be drained before servicing.

More About Sizing Pumps

Keen to size your own pump? Head to www.essentialrwh.com for a RWH pump-sizing example.

Chapter 8

Assuring End-Use Water Quality

GOOD UPFRONT DESIGN and sensible, pro-active maintenance of your system are the fundamental things that will have the biggest impact on your actual end-use water quality.

Although harvested rainwater may contain microbial, organic, and inorganic contaminants acquired from both atmospheric deposition and from leaching of materials (for instance roof surfaces, gutters, tanks, and other piping components), quality will improve significantly as rainwater moves through a well-designed — and well maintained — system. Despite some fear-mongering that exists out there, stored rainwater is usually of good quality and the risk of contracting illness from properly designed and serviced home-scale systems is actually very low.

For loads of fantastic research substantiating the above, see the Resources section and particularly the publications by Coombes (2016), enHealth (2012), Morrow (2012), and Morrow et al. (2007).

Summary of Good Design and Sensible Maintenance Practices

Throughout this book, we've emphasized good design and maintenance practices both for the system as a whole and for the individual components.

If you follow these recommendations:

i. Your RWH system will act as a treatment train.

ii. You will improve taste and aesthetics of your water.

iii. You will minimize health hazards and the potential contamination by harmful microorganisms.

iv. You will prevent problems that may result in needing corrective action (such as end-of-line sterilization or chlorine disinfection of your tank).

The design and maintenance practices recommended in this book are summarized in Table 8.1 (below).

Final Filtration and Water Testing

We only present final filtration now for a very important reason. Filtration, in of itself, is not a substitute for good design and sensible maintenance. If water-quality problems occur, the most likely cause is poor design; if not that, then it is usually the result of poor maintenance. Relying on filtration (or disinfection, for that matter) is treating the symptom and not the cause.

It's not to say that you shouldn't have final filtration. We emphasize that filtration should only play one small part in producing/storing quality water in a roof-to-tap RWH system.

Even if only low-quality water is required, final filtration is a good idea to safeguard your downstream plumbing against sedimentation and discoloration. For this purpose, a 0.5 micron filter is usually adequate.

For a higher-quality water supply, a 0.1–0.5 micron filter with activated carbon could work; it would remove organic and synthetic chemicals like pesticides. A carbon filter installed at the tap is a good idea because household plumbing (pipes, faucets, etc.) has been shown to degrade water quality. This would be the case whether you are using rainwater or a different water source, such as municipal water.

Table 8.1

Good Design	Sensible Maintenance
Collection • Select the least toxic or most benign roofing materials. • Enhance sterilization factors, reduce contamination vectors. • Slope gutters to properly drain.	**Collection** • Keep roof clean of debris. • Prune overhanging vegetation. • Frequently inspect your gutters and clean them out.
Pre-Filtration • Include a pre-filtration strategy in your design. • Place equipment in a location easy to access and inspect. • Include a downspout filter or equivalent upstream of the tank inlet.	**Pre-Filtration** • Check and clean screens or basket filters at a regular interval. • Repair screens as required. • Check first flush diverter regularly and clean out drainage port as required.
Conveyance Piping • Place PVC piping protected from direct sunlight, or in the shade, if possible. Consider painting or coating it to protect from UV degradation. Dry Systems: • Slope conveyance piping to ensure adequate drainage between rain events. Wet Systems: • Include a drainage port in the lowest point that is easily accessible.	**Conveyance Piping** • Replacing PVC piping before it gets old and starts to visibly degrade. Wet Systems: • Manually drain piping in between rain events or after long dry periods. • Include a downspout filter and mesh screens on inlets and outlets.
Storage • Follow the design directives to de-energize inlet water as much as possible. • Ensure your intake is at least 100 mm (4 in) above the bottom of the tank such as to not disturb the sludge layer. • Cut the inlet of the overflow at a 45 degree angle to ease "skimming" of floating particulate at the water surface. • Design overflow piping to prevent the possibility of backflow. • Screen all possible entry points for insects or other creatures. Below-grade • Ensure manway is properly sealed preventing surface water contamination. • Use correct below-grade rated water-tight fittings for all tank connections.	**Storage** • Avoid unnecessary sterilization, chlorination, or de-sludging of your tank. • Check and clean any tank inlet or outlet filters on a regular basis. • Repair screens as required. • Perform regular visual maintenance, both external and internal. • Check for structural integrity, make repairs as required.

Table 8.1: *Good design and sensible maintenance practices for roof-harvested home-scale RWH systems.*

If you have reason to be concerned about heavy metal contamination in your water, (e.g. you live close to an industrial area, coal plant, or other heavy manufacturing — or for any other reason), you may consider additional filtration at the drinking water tap. Although we've never used one, KDF filters have been recommended to us by numerous suppliers.

If you like to test your water (or if you are required to), your local or county health department can provide you with a list of laboratories. When collecting your water sample for testing, we recommend that you take the water sample from your end-use point, such as an indoor tap, and not directly from the tank, as is sometimes recommended. Testing tank water will only give a misleading result and a highly pessimistic view of your water quality.

From the *Guidance on the Use of Rainwater Tanks,* 3rd ed. (2012), Environmental Health Committee, Commonwealth of Australia:

Regular chemical or microbiological testing of domestic rainwater tanks is not needed. Microbial testing of rainwater from domestic tanks is rarely necessary and in most cases is not recommended. Water quality in rainwater tanks can change rapidly during wet weather and, during dry periods, the concentrations of indicator bacteria (E. coli) and faecal pathogen numbers decrease due to die-off (Edberg et al. 2000). Testing for special pathogens is often expensive and is generally only warranted as part of an outbreak investigation. If there are strong concerns about water quality, chlorination of tank water is a suitable alternative to testing. If microbial testing is undertaken, the parameter of choice is E. coli as an indicator of faecal contamination. Tests for total coliforms or heterotrophic plate counts are of little value as indicators of the safety of rainwater for drinking .(p. 26)

Regulatory Requirements for End-of-Line Treatment

We reiterate that the regulatory framework in North America runs the gamut from non-existent to completely obscure to over-the-top paranoid. As such, we can't provide a one-size-fits-all recommended approach, especially when it comes to ensuring that you meet any applicable local legal requirements.

In general, though, here is what we have noticed:

- Based on your end-use (irrigation, indoor, etc.), find out if your home-scale, private RWH system is regulated (it may or it may not be).

- If regulated, the regulator has likely divided RWH systems into two broad categories: potable and non-potable. Some regulators add additional granularity based on end-use (for instance CSA/ICC B805, has four end-use tiers based on potability and the potential of human contact).

- If your system does not meet the regulator's requirements for potability, you may be required to label all rainwater-supply fixtures and piping prominently, with something like: *Non-Potable — Do Not Drink.*

- If you plan on using rainwater for drinking or for indoor uses such as toilet flushing, you may have to add higher levels of filtration and disinfection than stated in the previous section, with drinking water requiring the highest levels of end-of-line treatment.

- The regulator's definition of end-of-line treatment likely provides the choice of ongoing chlorination (not recommended by the authors), or ozone or ultraviolet sterilization. Microfiltration (a filter with a 0.1–0.5 micron rating) might also be an option. In the US and Canada, any device purchased off-the-shelf will likely have to meet NSF 55 Class A (for drinking) or Class B (for other indoor uses).

- There may be also additional requirements for pH monitoring, management, and ongoing water quality testing.

Unfortunately, despite the fact that concerns raised in regard to rainwater quality have not been substantiated by scientific evidence (Spinks et al., as cited in Morrow, 2012), and that good design and sensible maintenance have been shown to deliver acceptable water quality for drinking and other uses (Coombes, 2016; enHealth, 2012; Morrow, 2012; Evans, et al., 2009; Morrow, et al., 2007) and that the health of rainwater consumers is the better or the same (Heyworth, 2006; Rodrigo, 2009) as non-rainwater consumers, there is a bias to mandate end-of-line sterilization or disinfection, adding (sometimes substantial) cost and complexity to the design.

Meeting Standard 63 and CSA/ICC B805

The ARCSA/ASPE/ANSI 63: Rainwater Catchment Systems (Standard 63), or the CSA/ICC B805 Rainwater Harvesting Systems are two standards that we think are most likely to be enacted by regulators. If you are designing a system to meet a published standard, the first thing you need to do is go and get a copy (easily available for purchase online) and read it. Next, ensure that your design (and operation) meets all of the required specifications.

Here we provide an example of end-of-line treatment system that could meet the filtration and disinfection requirements for potable drinking water as per Standard 63 and CSA/ICC B805. We provide manufacturer names and models here only to ease your research efforts and for illustration purposes, and not because we endorse one manufacturer over another. There are, of course, many other permutations and combinations and manufacturers that have products that could meet the end-of-line treatment requirements. Note also that in the design of a RWH system where good design and sensible maintenance practices are in place, we would lean towards meeting any final end-of-line treatment requirements using filtration or microfiltration (if allowed) versus using UV or other forms of disinfection.

The example end-of-line treatment system is shown in Figure 8.1 and the image is annotated with 12 numbers, representing different components as well as instrumentation and valves included to improve serviceability. Water from the rain tank and pump come in on the righthand side of the photo. Note that irrigation water bypasses this treatment system completely.

Bypass valve (number 1): This valve allows the entire treatment system to be bypassed. The valve stays closed except for exceptional circumstances. Note that local regulations may not allow you to even include one in the design.

Isolation valves (number 2, 12): These valves are in the open position unless servicing of the treatment system is required.

Pressure gauges (numbers 3, 5, 8, 10): Gauges are placed between all stages of filtration to help quickly diagnose if a filter is causing a pressure drop and a restriction in flow indicating that it's time to replace the filter.

Coarse particulate filter (number 4): This is a dual-gradient 25 to 1 micron filter inside of a standard filter housing. Having a coarse filter reduces the amount of servicing required on the second filtration stage. Filter Model: Pentair Pentek DGD-2501–20 Dual-Gradient Density Span Polypropylene Filter Cartridge. Housing Model: Big Blue Standard Filter Housing.

Fine particulate filter (number 6a, 6b): This ceramic filter provides micro filtration of bacteria, cysts, and particles down to .5 microns. It's cleanable and made of natural materials. Two filters are placed in parallel because of the flow rate limitations per unit of 19 liters/min (5 GPM). Filter Model: Doulton Rio 2000 Sterasyl Ceramic.

Activated Carbon Filter (number 7): This activated carbon filter, inside of a standard filter housing, removes odor, VOCs, and SOCs and improves taste. Filter Model: Pentair Pentek GAC — 20BB Granular Activated Carbon Cartridge. Housing Model: Big Blue Standard Filter Housing.

Ultraviolet Sterilization (number 9a, 9b): The sterilizer (9a) is a NSF Class A

certified device and capable of sterilizing up to 45 liters per minute (12 gpm). The controller is shown as 9b. Model: Viqua/Trojan UVMax D4 UV Sterilizer.

Check Valve (number 11): This valve ensures that water only flows in one direction, which is toward the indoor fixtures.

The cost of this system, installed in 2017, was about $3,000 plus approximately $1,000 for installation.

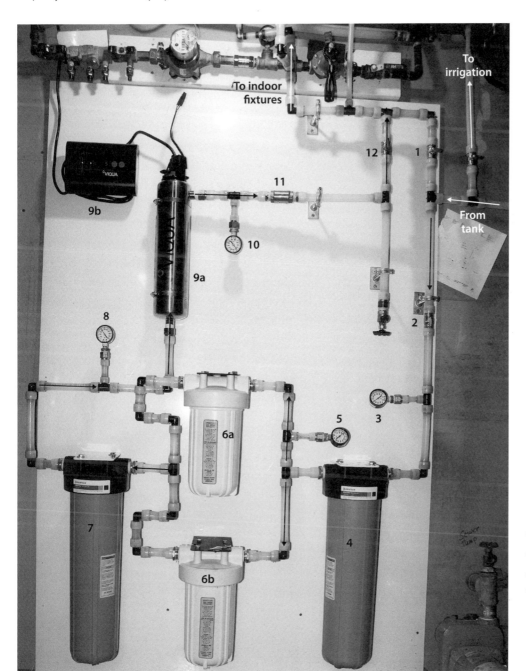

Fig. 8.1: *An example of a filtration and disinfection train that meets Standard 63 and CSA/ICC B805 for potable water use.*

CREDIT: VERGE PERMACULTURE

Rainwater pH

In chemistry, pH is a measure of acidity or alkalinity in a liquid solution, with 1 being extremely acidic and 14 being extremely alkaline. Rainwater tends to fall from the sky slightly acidic, with an average pH of about 5.5 in North America. This is completely natural; it's due to the naturally present carbon dioxide in the air dissolving into the rainwater. This slightly acidic pH is very different from an acidic pH created by contamination (such as sulfur pollution).

When it comes to pH and your rainwater, there are two important things to note:

• Rainwater pH will increase slightly (toward a neutral pH of 7) as the droplet of rain travels through a well-designed and sensibly maintained RWH treatment train (from the roof to the tap). (Coombs et al. 2006.).

• The pH values defined within drinking water guidelines are typically there to protect pipes from corrosion and are not stated as health guidelines.

Unless you live in a city with high pollution or near a pollution source such as a coal plant, we would not be concerned about making manual adjustments to the water pH in the rain tank, at least not for health reasons.

If you need to adjust pH over concerns of corrosion control, the document *Harvesting, Storing, and Treating Rainwater for Domestic Indoor Use* (TCEQ, 2007) is available free online and contains a recommended procedure for using baking soda to neutralize pH in the section on corrosion control. See the Resources section for a link.

Conclusion: The Importance of Good Design

OVER 3 MILLION AUSTRALIANS USE RAINWATER for drinking and 6.3 million use rainwater for some other household use. Virtually none of these systems use end-of-line sterilization or disinfection and there have been no epidemics from rainwater supply in Australia. In the last ten years, independent research (i.e. research not funded by water monopolies or product suppliers) has revealed that the quality of rainwater improves as a droplet of rain moves through a well-designed and sensibly maintained home-scale roof-harvested rainwater system (for example: Morrow 2012, Morrow et al. 2010; Evans 2010; Martin et al. 2010; Spinks 2007, Coombes 2002). This is, in part, because rainwater storages act as balanced ecosystems in a similar fashion to environmental systems that improve water quality (Evans et al, 2006). A well-designed RWH system includes multiple barriers: roof, leaf diverters, potentially a first flush diverter, tank inlet screens, and tank inlet/ outlet design that minimizes disturbances in the tank.

You must meet the legal requirements mandated by your local jurisdiction. But know that a fit-for-purpose RWH system that includes the good design and sensible maintenance practices outlined here will contribute far more towards ensuring water quality than relying purely on any prescriptive requirements for end-of-line treatment. As a homeowner, or rainwater harvesting practitioner, you need to make choices about the correct design for your particular situation. This book aims to highlight independent research that may help you to make informed choices.

Lastly, as we continue to learn, research, design, and operate RWH systems, we'll continue to add resources, new information and tools to our website: www.essentialrwh.com. Happy harvesting!

Appendix A: Conversion Table

Multiply	By	To Obtain
Centimeters	0.393701	Inches
Cubic foot	7.48052	Gallons
Cubic yards	202	Gallons
Cubic meter	1,000	Liters
Feet	30.48	Centimeters
Feet	0.3048	Meters
Feet of water	62.43	Pounds/square foot
Feet of water	0.434	Pounds/square inch
Gallons	3785	Cubic centimeters
Gallons	0.1337	Cubic feet
Gallons	3.785	Liters
Gallons water	8.3453	Pounds of water
Gallons/ft^2	40.7458	Liters/m^2
Gallons/minute	0.002228	Cubic feet/second
Gallons/minute	1440	Gallons/day
Gallons/minute	0.06308	Liters/second
Gallons/day	0.0006944	Gallons/minute
Gallons/day/square foot	1.604	Inches/day
Grams	0.002205	Pounds
Grams/liter	1,000	Parts/million

Hectares	2.471	Acres
Inches	2.54	Centimeters
Inches/day	0.6234	Gallons/day/square foot
Inches	25.4	Millimeters
kilometer/hr	0.621371	Miles/hr
Kilograms	2.205	Pounds
Liters	1,000	Cubic centimeters
Liters	0.03531	Cubic feet
Liters	1,000,000	Cubic millimeters
Liters	0.2642	Gallons, US
Liters/m²	0.02454	Gallons/ft²
Meters	3.281	Feet
Milligrams/liter	1	Parts/million
Millimeters	0.0394	Inches
Million gallons/day	1.54723	Cubic feet/second
Parts/million	8.345	Pounds/million gallons
Pounds	453.5024	Grams
Pounds of water	0.1198	Gallons
Pounds/square inch	2.31	Feet of water
Pounds/square inch	2.036	Inches of mercury
Temperature (°C)	1.8, then add 32	Temperature (°F)
Temperature (°F) minus 32	5/9	Temperature (°C)

Resources and References

For updates and clickable links, visit: www.essentialrwh.com

Books

The following books are all excellent resources for planning and designing landscape water-harvesting earthworks, using and designing greywater systems, and designing for landscape infiltration:

Hemenway, T. 2009. *Gaia's Garden.* Chelsea Green Publishing Company, White River Junction, VT.

Lancaster, B. 2010. *Rainwater Harvesting for Drylands and Beyond,* Volume 1 and Volume 2. Rainsource Press, Tucson, AZ.

Ludwig, A. 2007. *Create An Oasis With Greywater.* Oasis Design. Santa Barbara, CA.

Ludwig, A. 2013. *Water Storage, Tanks Cisterns, Aquifers, and Ponds.* Oasis Design. Santa Barbara, CA.

Mollison, B. 2012. *Permaculture: A Designers' Manual.* Tagari Publications. Tasmania, Australia.

Mollison, B. 2011. *Introduction to Permaculture.* Tagari Publications. Tasmania, Australia.

Reynolds, M. 1990. *Earthship, Volume II: Systems and Components.* Solar Survival Press. Taos, NM.

We also recommend this book for its excellent exploration of the concept of resilience:

Zolli, A. and Healy, A.M. 2013. *Resilience: Why Things Bounce Back.* Simon & Schuster.

Online Resources, Courses, and Consulting Information

Adaptive Habitat

Calgary, Alberta

Our consulting firm, Adaptive Habitat, offers a depth of practical expertise in building science and appropriate technology (solar, wind, combined heat and power), as well as rainwater harvesting, agro-ecology, ecosystem engineering, soil regeneration, and onsite wastewater treatment/septic design.
www.adaptivehabitat.ca

Harvesting Rainwater For Drylands and Beyond

Tucson, Arizona

Author Brad Lancaster maintains a very informative blog with many useful resources including materials and suppliers lists.
www.harvestingrainwater.com

Oasis Design

Santa Barbara, California

Author Art Ludwig and his team maintain another great online resource pertaining to rainwater and greywater.
http://oasisdesign.net

Urban Water Cycle Solutions

Australia

Dr. Coombes has spent more than 30 years dedicated to the development of systems understanding of the urban, rural, and natural water cycles with a view to finding optimum solutions for the sustainable use of ecosystem services, provision of infrastructure, and urban planning. His website is a wealth of information and research.
https://urbanwatercyclesolutions.com/

Verge Permaculture

Calgary, Alberta

Learn about our free resources (videos, blog, newsletter, and free online courses), as well as our premium courses on topics such as permaculture, land design, greenhouse design, and rainwater harvesting (in-person and online) at www.vergepermaculture.ca

For resources related directly to this book, including spreadsheet templates, rainfall data, RWH videos, optimization examples, the *Essential Rainwater Harvesting Tool* as well as links to relevant research and publications: www.essentialrwh.com

Climate Data

Here we've provided urls for US, Australian, and Canadian climate data. For other countries, and other ways to access data, use the tips provided in Chapter 3 and 5.

US Climate Data

The National Centers for Environmental Information (NOAA) publish Climate Normals here: https://www.ncdc.noaa.gov/data-access/land-based-station-data/land-based-datasets/climate-normals

Precipitation Frequency Data (used for determining the maximum design rainfall intensity) can be found here: https://hdsc.nws.noaa.gov/hdsc/pfds/ Select your state, then your weather station and scroll down to view the tabular or graphical precipitation frequency estimates, such as the 5-minute, 25-year event, or the 60-minute, 100-year event.

Australian Climate Data

For daily rainfall or monthly statistics, head to Australian Bureau of Meteorology *Climate Data Online*: http://www.bom.gov.au/climate/data/index.shtml

The Intensity Frequency Duration data (for determining the maximum design rainfall intensity) can be accessed here: bom.gov.au/water/designRainfalls/

Canadian Climate Data

The Canadian Climate Normals are published here: http://climate.weather.gc.ca/climate_normals/

To determine the maximum design rainfall intensity, download the Intensity-Duration-Frequency Files from the Engineering Climate Datasets webpage: http://climate.weather.gc.ca/prods_servs/engineering_e.html. Note that the Canadian government lists each location by number, which makes it very difficult to navigate the data. We've built an alphabetical key (for instance Edmonton = 3012205) to save significant time when searching for a location. See www.essentialrwh.com for access.

Publications and Research

Abbott, S., Caughley, B. 2012. "Roof-Collected Rainwater Consumption and Health." Paper presented at the 5th Pacific Water Conference. Nov 1, 2012. Auckland, NZ. Available online: pwwa.ws/pdfs/Stan_Abbott_RWH_Consumption_Health_Handout.pdf

American Water Works Association (AWWA). 2012. "Buried No Longer: Confronting America's Water Infrastructure Challenge." Available online: https://www.awwa.org/Portals/0/files/legreg/documents/BuriedNoLonger.pdf

Coombes, P.J., Barry, M., Smit, M. 2018. "Systems Analysis and Big Data Reveals Benefits of New Economy Solutions at Multiple Scales." Paper presented at WSUD 2018 and Hydropolis 2018. Feb 12–15, 2018. Perth, Western Australia.

Coombes, P.J. 2016. "Response to Reply by AR Ladson and MI Magyar," in Vol 19, No.1, pp. 88–90, *Australian Journal of Water Resources*, DOI. Available online: http://dx.doi.org/10.1080/13241583.2015.1131597

Coombes, P.J. 2015. Discussion on "Influence of Roofing Materials and Lead Flashing on Rainwater Tank Contamination by Metals," by M.I. Magyar, A.R. Ladson, C. Daipe, and V.G. Mitchell. *Australian Journal of Water Resources* 19(1).

Coombes, P.J., Barry, M.E. 2007. "The Effect of Selection of Time Steps and Average Assumptions on the Continuous Simulation of Rainwater Harvesting Strategies," *Water Science and Technology* Vol 55, No 4, pp. 125–133. IWA Publishing.

Coombes, P.J. et al. 2006. "Key Messages from a Decade of Water Quality Research into Roof Collected Rainwater Supplies." Paper presented at Hydropolis 2006. Perth, Western Australia.

Dean, J., Hunter, P.R. 2012. "Risk of Gastrointestinal Illness Associated with the Consumption of Rainwater: A Systematic Review." *Environmental Science and Technology* 46, pp. 2501–2507.

enHealth. 2010. "Guidance on the Use of Rainwater Tanks," 3rd Ed. Commonwealth of Australia. Available online: health.gov.au/internet/main/publishing.nsf/Content/ohp-enhealth-raintank-cnt.htm

Evans, C.A. et al. 2007. "Coliforms, Biofilms, Microbial Diversity and the Quality of Roof-Harvested Rainwater," *School of Environmental and Life Sciences* University of Newcastle, Callaghan NSW, Australia.

Evans, C.A. et al. 2009. "Extensive Bacterial Diversity Indicates the Potential Operation of a Dynamic Micro-Ecology Within Domestic Rainwater Storage Systems," *Science of the Total Environment* 407, pp. 5206–5215.

Jones, C. 2010. "Soil Carbon: Can It Save Agriculture's Bacon?" Paper presented at the Agriculture and Greenhouse Emissions Conference. Accessed 2018/01/15 (amazingcarbon.com/PDF/JONES-SoilCarbonandAgriculture.pdf (Revised May 18, 2010)

Harvie, J., Lent, T. (Draft 2011) "PVC-Free Pipe Purchasers' Report," *Healthy Building Network* Washington, DC. Accessed 2018/02/01, https://healthybuilding.net/uploads/files/pvc-free-pipe-purchasers-report.pdf

Heyworth J.S. et al. 2006. "Consumption of Untreated Tank Rainwater and Gastroenteritis Among Young Children in South Australia," *International Journal of Epidemiology* 35(4) pp. 1051–1058.

Lucas, S.A. et al. 2006. "Rainwater Harvesting: Revealing the Detail," *Water Journal of the Australian Water Association* 33. pp. 50–55.

Mack, E.A., Wrase, S. 2017. "A Burgeoning Crisis? A Nationwide Assessment of the Geography of Water Affordability in the United States," *PLoS ONE* 12(1): e0169488. https://doi.org/10.1371/journal.pone.0169488

Mechell, J. et al. 2010. "Rainwater Harvesting: System Planning." Texas AgriLife Extension Service. College Station.

Morrow, A. 2012. "Variations in Inorganic and Organic Composition of Roof-Harvested Rainwater: Studies at the Regional and Individual Site Level in Eastern and Southern Australia." Doctor of Philosophy thesis submitted to The University of Newcastle, Australia.

Morrow, A. et al. 2007. "Elements in Tank Water — Comparisons with Mains Water and Effects of Locality and Roofing Materials." Rainwater and Urban Design 07 — Incorporating the 13 International Rainwater Catchment Systems Conference and 5th Water Sensitive Urban Design Conference, August 21–23, Sydney, Australia.

Regional District of Nanaimo, *Rainwater Harvesting Best Practices Guidebook*. Free download at www.rdn.bc.ca/rainwater-harvesting.

Rodrigo S. et al. 2009. "Drinking Rainwater: A Double-Blinded, Randomized Controlled Study of Water Treatment Filters and Gastroenteritis Incidence," *American Journal of Public Health* 101(5) pp. 842–847.

Spinks, A. 2007. "Water Quality, Incidental Treatment, Train Mechanisms and Health Risks associated with Urban Rainwater Harvesting Systems in Australia." Doctor of Philosophy thesis submitted to The University of Newcastle, Australia.

TCEQ. 2007. "Harvesting, Storing, and Treating Rainwater for Domestic Indoor Use." Texas Commission on Environmental Quality (TCEQ). GI- 366. Jan 2007. Available online: http://rainwaterharvesting.tamu.edu/files/2011/05/gi-366_2021994.pdf

USDA National Engineering Handbook. 1997. "Irrigation Guide, Part 652." Available online: https://directives.sc.egov.usda.gov/OpenNonWebContent.aspx?content=17837.wba

Codes and Standards

There are many codes and standards that are applicable to rainwater harvesting systems. Here is a listing of some of the most commonly used codes and standards in RWH systems in North America.

American Society of Plumbing Engineers
- ARCSA/ASPE/ANSI 63: Rainwater Catchment Systems

ASTM International
- ASTM E2727–2010e1: Standard Practice for Assessment of Rainwater Quality

American Water Works Association (AWWA)
- AWWA D107: Composite Elevated Tanks for Water Storage
- AWWA D100: Welded Carbon Steel Tanks for Water Storage

CSA Group
- CSA/ICC B805–2015 Draft. Rainwater Harvesting Systems
- CAN/CSA-B64.10–11/CAN/CSA-B64.10.1–11: Manual for the selection and installation of backflow prevention devices/Manual for the maintenance and field testing of backflow prevention devices
- CSA B126–13: Water cisterns

Health Canada
- Guidelines for Canadian Drinking Water Quality (published by Health Canada on behalf of the Federal-Provincial-Territorial Committee on Drinking Water)

International Association of Plumbing and Mechanical Officials (IAPMO)
- IAPMO/ANSI Z1002: Rainwater Harvesting Tank
- IAPMO. 2018. Uniform Plumbing Code. Available online at http://epubs.iapmo.org/2018/UPC

National Research Council (NRC)
- National Building Code of Canada 2010
- National Plumbing Code of Canada 2010

NSF International
- NSF 53–2013: Drinking Water Treatment Units — Health Effects
- NSF 55–2014: Ultraviolet Treatment
- NSF 60: Drinking Water Treatment Chemicals — Health Effects
- NSF 61–2014a: Drinking Water System Components — Health Effects
- NSF 372–2011: Drinking Water System Components — Lead Content
- NSF P151–1995: Health Effects from Rainwater Catchment Systems

Index

About the Authors

ROB AND MICHELLE left the oil and gas industry in 2008 and launched Verge Permaculture, now an internationally recognized and award-winning education company. Through Verge they are leading the next wave of permaculture education and teaching career-changing professionals how to build successful businesses combining technology with earth science and eco-entrepreneurism.

They also own and operate Adaptive Habitat, a unique and leading-edge property design and management firm through which they leverage their 20 years of combined experience in engineering, project management, ecological design, and sustainable technologies. As skilled Professional Engineers, they offer a depth of practical expertise in building science and appropriate technology (solar, wind, combined heat and power), as well as rainwater harvesting, agro-ecology, ecosystem engineering, soil regeneration, and onsite wastewater treatment/septic design.

With ongoing training and practice in dozens of sustainable technologies ranging from natural building materials to greywater systems, wind turbines to soil and wetland restoration, they've converted their urban home in Calgary, Alberta, Canada, into a living permaculture project complete with a front-yard food forest, water-harvesting features, passive solar greenhouse, and energy retrofits on the house.

Since launching Verge, they've helped more than 5,000 students and a growing number of clients design and/or create integrated systems for shelter, energy, water, waste, and food, all while supporting local economy and regenerating the land.

In keeping with Verge's mission to grow the permaculture groundswell, Rob donates much time as a popular and passionate public speaker for community events, nonprofit groups, and businesses. He has been invited to lecture at local colleges and universities, and to speak at both regional and international conferences.

The two have earned testimonials from bestselling author Toby Hemenway, dryland-restoration guru Craig Sponholtz, and "beyond-organic farming" superstar Joel Salatin among others. Leading permaculture designer/trainer Geoff Lawton calls Verge "one of North America's premier permaculture design and education companies."

Rob finds balance in his life by spending free time outdoors with Michelle and their two children, and by devoting time to his meditation practice. In his odd extra moments, he can be found tinkering in his shop, cooking up a new recipe in the kitchen, picking out chords on his guitar … or jotting down notes on his latest entrepreneurial idea.

Michelle is focusing her time and energy primarily on the couple's two young children, while growing microgreens, sprouts, herbs, and veggies in the Verge greenhouse and gardens. An avid swimmer, she also enjoys long-distance cycling in the summer and cross-country skiing in the winter.

A Note About the Publisher

NEW SOCIETY PUBLISHERS is an activist, solutions-oriented publisher focused on publishing books for a world of change. Our books offer tips, tools, and insights from leading experts in sustainable building, homesteading, climate change, environment, conscientious commerce, renewable energy, and more — positive solutions for troubled times.

We're proud to hold to the highest environmental and social standards of any publisher in North America. This is why some of our books might cost a little more. We think it's worth it!

- We print all our books in North America, never overseas
- All our books are printed on **100% post-consumer recycled paper**, processed chlorine free, with low-VOC vegetable-based inks (since 2002)
- Our corporate structure is an innovative employee shareholder agreement, so we're one-third employee-owned (since 2015)
- We're carbon-neutral (since 2006)
- We're certified as a B Corporation (since 2016)

At New Society Publishers, we care deeply about *what* we publish — but also about how we do business.

Download our catalogue at https://newsociety.com/Our-Catalog or for a printed copy please email info@newsocietypub.com or call 1-800-567-6772 ext 111

New Society Publishers

ENVIRONMENTAL BENEFITS STATEMENT

For every 5,000 books printed, New Society saves the following resources:[1]

39	Trees
3,549	Pounds of Solid Waste
3,904	Gallons of Water
5,093	Kilowatt Hours of Electricity
6,451	Pounds of Greenhouse Gases
28	Pounds of HAPs, VOCs, and AOX Combined
10	Cubic Yards of Landfill Space

[1]Environmental benefits are calculated based on research done by the Environmental Defense Fund and other members of the Paper Task Force who study the environmental impacts of the paper industry.